T0134546

EcoProduction

Environmental Issues in Logistics and Manufacturing

Series Editor

Paulina Golinska-Dawson, Poznań, Poland

The EcoProduction Series is a forum for presenting emerging environmental issues in Logistics and Manufacturing. Its main objective is a multidisciplinary approach to link the scientific activities in various manufacturing and logistics fields with the sustainability research. It encompasses topical monographs and selected conference proceedings, authored or edited by leading experts as well as by promising young scientists. The Series aims to provide the impulse for new ideas by reporting on the state-of-the-art and motivating for the future development of sustainable manufacturing systems, environmentally conscious operations management and reverse or closed loop logistics.

It aims to bring together academic, industry and government personnel from various countries to present and discuss the challenges for implementation of sustainable policy in the field of production and logistics.

More information about this series at http://www.springer.com/series/10152

Marjana Petrović · Luka Novačko
Editors

Transformation of Transportation

Editors
Marjana Petrović
Department of Railway Transport
Faculty of Transport and Traffic Sciences
University of Zagreb, Zagreb, Croatia

Luka Novačko
Department of Road Transport
Faculty of Transport and Traffic Sciences
University of Zagreb, Zagreb, Croatia

ISSN 2193-4614 ISSN 2193-4622 (electronic)
EcoProduction
ISBN 978-3-030-66466-4 ISBN 978-3-030-66464-0 (eBook)
https://doi.org/10.1007/978-3-030-66464-0

This Springer imprint is published by the registered company Springer Nature Switzerland AG
The registered company address is: Gewerbestrasse 11, 6330 Cham, Switzerland

Preface

This monograph presents the original scientific manuscripts submitted for publication at the International Conference—The Science and Development of Transport (ZIRP 2020), organized by the University of Zagreb, Faculty of Transport and Traffic Sciences, Zagreb, Croatia. Co-organizers of the ZIRP 2020 conference are Wrocław University of Science and Technology, University of Bremen, International Road Federation, University of Zagreb Faculty of Mechanical Engineering and Naval Architecture, University Politehnica of Bucharest, UNESCO Chair of Engineering for Society, University of Zagreb Faculty of Economics & Business, European Platform of Transport Science, Croatian Academy of Engineering and AMAC-FSC. The conference was held online, from September 29 to 30, 2020, aiming to bring together the scientists and the practitioners by putting forward innovative solutions available to everyone. This monograph presents the newest scientific research, case studies and best practices in transport and logistics. The monograph will be of interest for experienced researchers, professionals as well as Ph.D. students in the field of transport and logistics. We would like to express our gratitude to the authors that contributed to this monograph.

In the paper "Effect of Different Stop Sign Configurations on Driving Speed When Approaching a Rural Intersection at Night-Time," the research aims to evaluate how enhanced visibility stop signs (fluorescent signs) affect the driving speed when approaching a rural intersection at nighttime. The research was conducted using a driving simulator.

The next paper "Traffic Flow Simulators with Connected and Autonomous Vehicles: A Short Review" pairs the three most common validated traffic simulators VISSIM, AIMSUN and SUMO with network simulators NS-3 and OMNET++ to model AVs and CAVs in a simulation environment and compares their features.

The following paper "Application of Dimensionless Method to Estimate Traffic Delays at Stop-Controlled T-Intersections" discusses current research on developing a new model of estimating traffic delays experienced by the minor approach vehicles at stop-controlled T-intersections.

"In-Depth Evaluation of Reinforcement Learning Based Adaptive Traffic Signal Control Using TSCLAB" applies an augmented version of the previously developed MATLAB-based tool Traffic Signal Control Laboratory (TSCLab) to evaluate a newly proposed ATSC based on self-organizing maps and reinforcement learning. Its performance is evaluated using appropriately chosen measures of effectiveness obtained in real time using a VISSIM-based microscopic simulation environment and a realistic traffic scenario.

The paper "Discrete Simulation Model for Urban Passenger Terminals" develops a simulation model for a hypothetical passenger terminal using the main Romanian train station's topology. The obtained data can be used to optimize the number of access gates, the stairs, the waiting area, etc.

The aim of the paper "Characteristics of Departing Passenger Reports to the Passport Control Queuing System" is to present the research results concerning the analysis of the stream of passenger reports to the queuing system of passport control in the departure hall.

The next paper "Situation in Railway Sidings Operation in Slovakia Based on the Selected Criteria" focuses on examining the railway siding performance share in the overall rail transport performance. It provides a dependence analysis of railway siding performance on selected criteria: total transport volume in international transport, transport price and the number of railway siding services.

The following paper "Applying Multi Criteria Analysis in Evaluation of Distribution Channels" outlines the possibilities of applying the AHP method of multi-criteria analysis in evaluating optional distribution channels in the distribution of confectionery products.

The object of the paper entitled "Development Barriers of Eurasian Container Transportation" is container transportation of goods through the countries of the Eurasian Eco-nomic Union (EAEU), the People's Republic of China (PRC), and the European Union (EU). The relevance of this topic is due to the existence of certain barriers and the need to overcome them for the development of cargo turnover in this direction.

The purpose of the next paper "Airline Fleet Rotables Staggered Replacement Scheduling Using Dynamic Approach" is to describe the application of a dynamic approach for scheduling airline rotables preventive maintenance to minimize earliness costs and maximize on wing time of the components.

The paper "Monitoring Traffic Air Pollution Using Unmanned Aerial Systems" introduces unmanned aerial vehicles (UAVs) equipped with various air quality sensors offering new approaches and capabilities for monitoring air pollution, as well as studying atmospheric trends, such as climate change while ensuring safety in urban and industrial areas.

The authors of "Instruments for Career Development in the Air Transport Industry" propose the following tools for supporting the employment, carrier development and professional orientation in air transport: two methodology developments within a European project and online platforms providing information on employment opportunities and qualifications in aviation.

The next paper "Drivers of Change for Smart Occupations and Qualifications in Aviation" focuses on the challenges and changes in future occupations and identifies new qualifications needed to meet the new air transport trends.

The final paper "A Framework to Understand Current and Future Competences and Occupations in the Aviation Sector" presents an analysis of occupations and competencies required for current and emerging roles in the aviation sector. A mixed-method approach was employed, which combined desk studies and the involvement of external aviation stakeholders.

Zagreb, Croatia

Marjana Petrović
Luka Novačko

Contents

Effect of Different Stop Sign Configurations on Driving Speed When Approaching a Rural Intersection at Night-Time

Dario Babić, Darko Babić, Mario Fiolić, and Marko Ružić

Abstract As part of the road network utilized by all road users, intersections are places of high complexity and conflict risks. Statistics show that 40–60% of all road accidents occur at intersections, while around 20% of them result in fatalities. The consequences of collisions at intersections are particularly severe on rural roads during night-time due to higher speeds than in urban areas and poor visibility. Therefore, the aim of this study is to investigate how enhanced-visibility stop signs (fluorescent signs) affect driving speed when approaching a rural intersection at night-time. The study was conducted using a driving simulator comprising a 6.61 km rural road with six intersections. The results show that additional stop signs influence driving behavior and encourage drivers to reduce speed when approaching a rural intersection at night. This particularly relates to signs with enhanced visibility (fluorescent signs). The results of the study could be useful for road engineers and authorities, especially in developing countries, to increase road safety at dangerous unsignalized rural intersections by implementing low-cost traffic control measures.

Keywords Traffic signs · Unsignalized intersection · Fluorescent traffic signs · Road safety · Driving speed

D. Babić (✉) · D. Babić · M. Fiolić · M. Ružić
Faculty of Transport and Traffic Sciences, University of Zagreb, Zagreb, Croatia
e-mail: dario.babic@fpz.unizg.hr

D. Babić
e-mail: darko.babic@fpz.unizg.hr

M. Fiolić
e-mail: mario.fiolic@fpz.unizg.hr

M. Ružić
e-mail: marko.ruzic@fpz.unizg.hr

M. Petrović and L. Novačko (eds.), *Transformation of Transportation*, EcoProduction,
https://doi.org/10.1007/978-3-030-66464-0_1

1 Introduction

Road accidents are one of the leading causes of death worldwide (World Health Organization 2018). Although road safety is improving in most countries, progress remains slow due to the dynamic and complex nature of road traffic and the interconnectedness of factors related to the roadway and its environment, the vehicle, and the human. Statistics show that most accidents (54%) in the EU occur on rural roads although traffic volumes are much lower compared to urban areas (European Commission 2019). Furthermore, intersection collisions account for 40–60% of all road accidents (European Commission 2018a). This is because, intersections are in most cases, part of the road network used by all road users (cars, trucks, motorcycles, bicycles, pedestrians), and as such, are places of high complexity and conflict risks. In the EU, around 20% of traffic accidents that occur at intersections are fatal (European Commission 2018b). In contrast, in the US, on average a of quarter of road fatalities and roughly half of all injuries occur at intersections (Federal Highway Administration 2019).

From a safety perspective, there are several causes of intersection collisions, namely high approach speed, improper speed control, insufficient sight distance to oncoming vehicles, lack of intersection visibility (road users do not perceive the intersection), lack of gaps in traffic, complex intersection layout and poor road surface condition (Biancardo et al. 2019; Yang et al. 2019; Himes et al. 2018).

This shows that visibility is a major factor in maintaining safety at intersections. While driving, drivers receive more than 90% of information visually (Gregersen and Bjurulf 1996), so the timeliness of the information they receive is critical for appropriate response. At night, the human field of vision is narrowed and shortened, and the perception of colour, shape, texture, contrast and movement is reduced (Plainis et al. 2006), which impairs the driver's ability to avoid collisions and thus increases the overall risk of accidents (Rice et al. 2009; Sullivan and Flannagan 2002; Li et al. 2018).

In order to achieve satisfactory level of traffic flow and safety, most countries use different types of traffic control at intersections, namely uncontrolled intersections, intersections controlled by traffic signals, the use of a right-of-way sign on the minor road, the use of a stop sign on the minor road, and the use of stop signs on both minor and major roads (Polus 1985). The decision as to which of these or other alternatives should be used depends on traffic safety, road type and function, number of merging routes, traffic volume and type, design and operating speed, priority setting, terrain, available space, environmental concerns, cost, etc. The relative importance of each factor varies from case to case and should be taken into account (World Road Association 2003).

When looking at efficient low-cost solutions, traffic signs and road markings have proven to be the most cost-effective measures (Hummer et al. 2010; Thurston 2009; Hallmark et al. 2012; McGee and Hanscom 2006). Although most rural intersections are regulated by right-of-way or stop signs, drivers often perceive them too late and thus do not adjust their driving speed in a timely and proper manner to the upcoming situation. This is mainly due to the inadequate quality of the signs, their lack of maintenance and drivers' familiarity with the route. The quality of the sign is mainly determined by its retroreflection and chromaticity properties. The studies have shown that signs which do not meet the minimum prescribed values and are improperly maintained may contribute to traffic accidents, especially at night-time (Xu et al. 2018; Šarić et al. 2018; Ferko et al. 2019). Also, several studies show that route familiarity affects driver perception of different road and safety elements, including traffic signs, and that drivers, especially older and more experienced ones, perceive fewer signs as they become more familiar with the route (Babić 2017; Yanko and Spalek 2013).

From all the above, it can be concluded that rural road intersections at night time are a high collision risk location. Therefore, this study aims to evaluate how enhanced-visibility stop signs (fluorescent signs), as a low-cost traffic-control measure, affect driving speed when approaching a rural intersection at night-time. The study was conducted using a driving simulator in which a 6.61 km long rural road with six identical, but differently controlled intersections has been created. The results of the study could be useful for road engineers and authorities to increase road safety at dangerous unsignalized rural intersections through low-cost traffic control measures.

2 Methodology

2.1 Research Equipment

For the study, we used the Carnetsoft B. V. driving simulator (Fig. 1) consisting of a driver section (driver's seat with pedals, steering wheel, and shifter) and three interconnected displays, 30' in size, 5760 × 1080 resolution, and 30 Hz refresh rate. The hardware consisted of a computer with NVidia GeForce GTX 1080 Ti graphics processing unit (GPU) and 3 GB of video memory, Intel Core i7 7700 K central processor unit (CPU) with four cores, eight threads and frequency of 4.20 GHz, 32 GB of RAM, 250 GB SSD for storage and Windows 10 Pro 64-bit operating system. The simulator provides an interactive representation of reality with a 210° environment with over six channels (left, centre, and right views plus three rearview mirrors).

The described simulator has been used in several studies related to driver behaviour, which validates its use in this study (van Winsum 2018, 2019a, b).

2.2 Scenario Design

The scenario simulated night-time driving on a two-way rural road with 3.25 m wide roadway lanes and with active traffic in the opposite direction. The road section

Fig. 1 Carnetsoft B. V. driving simulator used in the study

was 6.61 km long and included six four-way intersections with traffic from other directions. Drivers had the right of way at three intersections (RW 1, RW 2, and RW 3), but not at the other three (Stop 1, Stop 2 and Stop 3). At one of these three intersections, traffic was controlled by a stop sign placed at the intersection according to Croatian standards (Stop 1). At the second, in addition to the stop sign at the intersection, an additional sign was placed 200 m before the intersection (Stop 2) with a supplementary plate defining the distance to the intersection. The third controlled intersection had in total three stop signs: the first one was placed on the right side of the roadway, 200 m before the intersection with an additional plate defining the distance to the intersection, while the second two were fluorescent stop signs located at the intersection on both the right and the left side of the roadway (Stop 3). The design of the intersections is presented in Fig. 2.

The roadway was marked with 15 cm wide white edge and centre lines according to Croatian design standards. The speed limit was set between 50 and 90 km/h. The scenario also included houses and other environmental elements such as trees. Both the sound of traffic in the environment and the sound of the participant's car were included. The summary of the scenario configuration is presented in Table 1.

Fig. 2 Different configurations of stop signs in the scenario

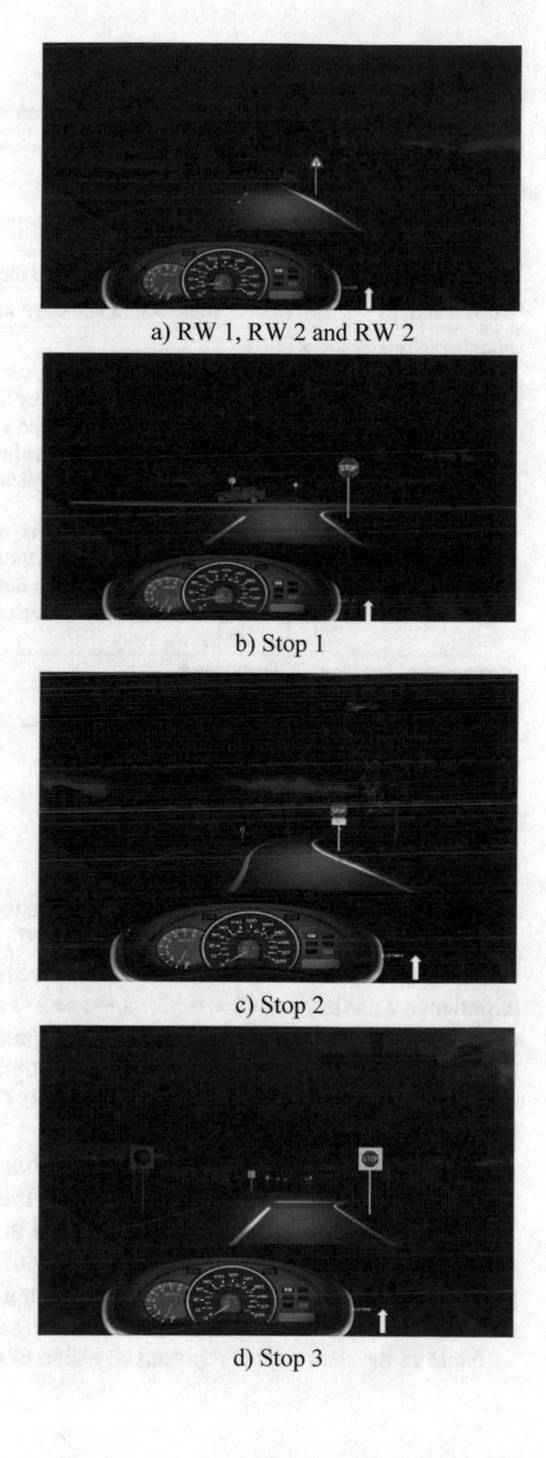

a) RW 1, RW 2 and RW 2

b) Stop 1

c) Stop 2

d) Stop 3

Table 1 Summary of scenario design

Parameter	Description
Length	6.61 km
Conditions	Night-time
Road type and width	Two-way rural road
Road lane width	3.25 m
Traffic signs	Active; placed according to the Croatian standard
Road markings	15 cm white centre and edge lines
Number of intersections	6 – 3 with the right of way (RW 1, RW 2, and RW 3) – 3 with traffic controlled by a stop sign: (a) Stop 1—stop sign placed at the intersection (b) Stop 2—stop sign at the intersection + stop sign 200 m before the intersection with a supplementary plate defining the distance to the intersection (c) Stop 3—three stop signs: one placed on the right side of the roadway 200 m before the intersection with an additional plate defining the distance to the intersection, and two fluorescent stop signs located at the intersection on both the right and the left side of the roadway
Speed limit	50–90 km/h
Sound	Active
Traffic	Active

2.3 Participants

A total of 32 volunteers with valid driving license were recruited, of which 23 were male (71.88%), and nine were female (28.13%). The mean age of the participants was 25 years ($\bar{x} = 25.11$; range = 21.6–29.8; SD = 1.81) while their mean driving experience was six years ($\bar{x} = 6.53$; range = 2–11; SD = 2.18). Participant selection criteria were based on the fact that young, predominantly male drivers are more likely to be involved in road accidents, although they tend to drive less frequently than older drivers (Bener and Crundall 2008; Gray et al. 2008; Scott-Parker and Oviedo-Trespalacios 2017).

Participants rated their driving ability with a mean score of 8.22 (on a scale of 1-10), and most of them (40.62%) reported that they were active drivers. Out of 32 participants, nine (28.12%) were involved in a traffic accident as a driver (they were involved in ten accidents in total), and 40% of them caused the accident. Six participants have mild refractive error and wear glasses or lenses while driving.None of the participants reported any signs of driving simulator sickness.

None of the participants reported any sign of driving simulator sickness.

2.4 Procedure

The testing room was set up in the Department of Traffic Signaling, Faculty of Transport and Traffic Sciences, University of Zagreb, Croatia. Before conducting the test, each participant was familiarized with the research equipment and the research procedure. The researchers instructed the participants that their driving skills and abilities would not be assessed and that they were free to drop out at any time, especially if they experienced side effects such as simulator sickness. Participants also signed an informed consent form to participate and completed a short questionnaire related to personal information such as age, gender, date of obtaining a driver's license, assessment of their driving ability, number of traffic accidents they were involved in and caused, frequency of driving, and other comments and possible problems related to their visual system. The overall aim of the study was not revealed to the participants in order to avoid bias in the results.

Prior to test driving, participants had a warm-up period of approximately 5-10 minutes to familiarize themselves with the driving simulator.

2.5 Data Analysis

To validate the effect of different stop sign configurations on driving speed when approaching rural intersections at night-time, driving speed was recorded at four locations before each intersection, namely 300, 200, 100, and 150 m. The measurement points were established based on the results of previous studies that used a similar methodology to evaluate the effectiveness of different perception measures to increase road safety (Ariën et al. 2017; Hussain et al. 2019; Montella et al. 2015).

The above data were extracted from the Carnetsoft B. V. "Data Analysis" software and analysed using univariate ANOVA.

3 Results

First, the interaction between driving speed and intersection type, measurement points, and participants' gender and driving experience was tested. Participants' age was excluded as a variable for two reasons: (1) the range was small (21.6–29.8 years with $SD = 1.81$ years); (2) participant age was highly correlated with driving experience (Spearman = 0.867).

The results of the ANOVA analysis show that driving speed differs significantly between intersections and measurement points ($p < 0.05$) while gender and driving experience do not ($p = 0.342$ and $p = 0.439$, respectively).

Table 2 Results of multiple comparisons between driving speed and intersections

(I) Intersection	(J) Intersection	Mean difference (I – J)	p	95% Confidence interval	
				Lower bound	Upper bound
RW 1	RW 2	−1.5599	0.072	−3.2603	0.1405
	RW 3	−1.4183	0.102	−3.1187	0.2821
	Stop 1	2.8289	0.001	1.1285	4.5293
	Stop 2	6.2186	0.000	4.5182	7.9190
	Stop 3	15.4176	0.000	13.7172	17.1180
RW 2	RW 3	0.1416	0.870	−1.5588	1.8420
	Stop 1	4.3888	0.000	2.6884	6.0892
	Stop 2	7.7785	0.000	6.0781	9.4789
	Stop 3	16.9775	0.000	15.2771	18.6779
RW 3	Stop 1	4.2472	0.000	2.5468	5.9476
	Stop 2	7.6369	0.000	5.9365	9.3373
	Stop 3	16.8360	0.000	15.1356	18.5364
Stop 1	Stop 2	3.3897	0.000	1.6893	5.0901
	Stop 3	12.5888	0.000	10.8884	14.2892
Stop 2	Stop 3	9.1990	0.000	7.4986	10.8994

Second, we analysed the difference in speed for each intersection. We conducted a post-hoc test using the LSD test (the significance level was set at 0.05). The results are presented in Table 2.

The results in Table 1 show that there was no statistical difference in driving speed ($p > 0.05$) between all three intersections where participants had the right of way (RW 1, RW 2, and RW 3). However, driving speeds at these intersections were statistically higher ($p < 0.05$) than at the intersections where traffic was controlled with stop signs (Stop 1, Stop 2, and Stop 3), which was expected.

A statistical difference ($p < 0.05$) in driving speed was also found between each controlled intersection (Stop 1, Stop 2, and Stop 3). The driving speed was lowest at the intersection "Stop 3" which was controlled with three stop signs: a regular stop sign placed on the right side of the roadway 200 m before the intersection with an additional sign defining the distance to the intersection, and two fluorescent stop signs located at the intersection on both the right and left sides of the roadway. The intersection that was controlled with two stop signs (Stop 2): one at the intersection and one with additional sign defining the distance to the intersection placed 200 m before the intersection, had an average speed 9.2 km/h higher compared to the intersection "Stop 3". The intersection with only one stop sign (Stop 1) had, on average, 3.4 km/h higher driving speed compared to the intersection "Stop 2" and 12.6 km/h to intersection "Stop 3". The mean driving speed for each intersection is shown in Fig. 3.

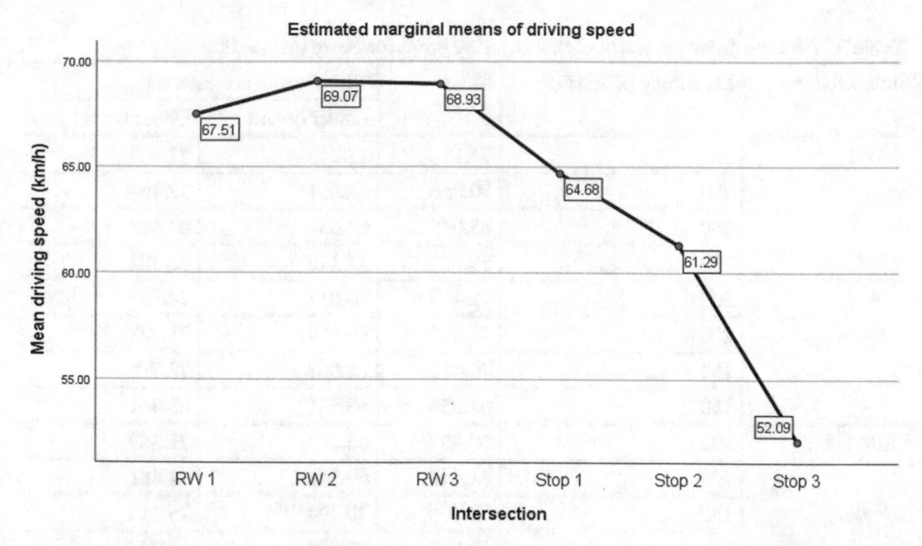

Fig. 3 The mean driving speed for each intersection

In addition, we analysed how driving speed changed at each measurement point (300, 200, 100, and 150 m before the intersection) depending on the type of intersection. From Table 3, one can conclude that the average driving speed at the same measurement points differs to some extent between intersections with right-of-way (RW1, RW 2, and RW 3) and the intersection with a stop sign (Stop 1). In contrast, a continuous decrease in speed at the same measurement points was observed at the controlled intersections (Stop 2 and Stop 3).

4 Discussion and Conclusion

Due to the interactions between the different road users and thus the overall complexity of the situation, intersections have a high accident risk. Statistics show that around 40–60% of all road accidents (depending on the country) occur at intersections, of which around 20% are fatal. Due to higher driving speeds, the consequences of collisions at rural intersections are particularly severe. In addition, the risk and severity of accidents increase at night (Rice et al. 2009; Sullivan and Flannagan 2002; Li et al. 2018).

For these reasons, this study aimed to evaluate how enhanced-visibility stop signs (fluorescent signs) affect driving speed when approaching a rural intersection at nighttime. To achieve the aim of the study, a scenario simulating a 6.61 km long rural road with six intersections has been created. Drivers had the right of way on three intersections while on the other three the traffic was controlled with a stop sign. The first intersection had a stop sign. The second one had a stop sign at the intersection and an additional sign 200 m before the intersection with a supplementary plate

Table 3 Average driving speed at each measuring point for every intersection

Intersection	Measuring point (m)	Mean	95% Confidence interval	
			Lower bound	Upper bound
RW 1	300	75.158	72.753	77.563
	200	70.056	67.651	72.460
	150	65.039	62.634	67.443
	50	59.777	57.372	62.182
RW 2	300	72.418	70.013	74.823
	200	72.323	69.919	74.728
	150	70.471	68.067	72.876
	50	61.056	58.651	63.461
RW 3	300	68.954	66.549	71.359
	200	71.946	69.542	74.351
	150	72.709	70.304	75.114
	50	62.093	59.688	64.498
Stop 1	300	66.194	63.789	68.599
	200	70.285	67.880	72.690
	150	69.489	67.084	71.893
	50	52.746	50.342	55.151
Stop 2	300	71.674	69.269	74.078
	200	64.874	62.470	67.279
	150	58.715	56.310	61.120
	50	49.892	47.487	52.296
Stop 3	300	60.097	57.692	62.502
	200	51.835	48.930	53.739
	150	51.313	49.408	54.218
	50	45.114	42.709	47.519

defining the distance to the intersection. The third controlled intersection had in total three stop signs: regular stop sign placed on the right side of the roadway, 200 m before the intersection with additional plate defining the distance to the intersection, and other two were fluorescent stop signs located at the intersection on both the right and the left side of the roadway.

The results show that participants did not significantly change their driving speed when approaching an intersection where they had the right of way. Driving speed for this intersection ranged from 67.5 km/h (RW 1) to 69.1 km/h (RW 2). On the other hand, as expected, driving speed at controlled intersections was lower than at intersections with right-of-way, ranging from 52.1 km/h (Stop 3) to 54.7 km/h (Stop 1).

Further analysis confirmed a statistical difference in driving speed between each controlled intersection. Comparison of the three controlled intersections revealed that driving speed was lowest at the intersection "Stop 3" controlled with three stop signs: a regular stop sign located 200 m before the intersection on the right side of the roadway, with an additional plate defining the distance to the intersection, and two fluorescent stop signs located at the intersection on both the right and left sides of the roadway. The intersection controlled by two stop signs (Stop 2) had an average speed 9.2 km/h higher compared to the "Stop 3" intersection. The intersection with only one stop sign (Stop 1) had an average of 3.4 km/h higher driving speed compared to the "Stop 2" intersection and 12.6 km/h higher driving speed compared to the "Stop 3" intersection.

Analysis of changes in driving speed at each measurement point (300, 200, 100 and 150 m before the intersection) by type of intersection suggests that there is some difference in average driving speed at the same measurement points for both intersections with right-of-way and intersection with a stop sign. This could be explained by the fact that participants did not know that a controlled intersection was ahead of them until they recognised the stop sign , which according to (Priambodo and Siregar 2018) is about 70 m on average. At the other two controlled intersections, driving speed decreased continuously at each measurement point because an additional stop sign was placed 200 m before the intersection.

The results confirm that additional stop signs affect driver behaviour and encourage drivers to reduce speed when approaching a rural intersection at night. This especially refers to enhanced-visibility signs (fluorescent signs). Namely, fluorescent is not a natural colour and stands out from the environment. For this reason, signs of that colour attract the driver's eye more and have greater detection and legibility distance compared to non-fluorescent signs (Zwahlen and Schnell 1977; Schnell et al. 2001; Burns and Johnson 1997). Also, previous studies suggest that a more active eye, to some extent, means greater attentional functions and better overall driving performance (Mackenzie and Harris 2017; Anstey et al. 2012).

This study confirms the general findings from previous studies (Hummer et al. 2010; Hallmark et al. 2012; Montella et al. 2015) that traffic signs represent one of the most cost-effective solutions for increasing traffic safety. Overall, this study could be useful to road engineers and authorities, especially in developing countries, in order to increase road safety on dangerous unsignalized rural intersections by implementing low-cost traffic control measures.

Although this study provided valuable results, it has certain limitations. They are primarily related to the number and age distribution of participants. All participants in the study were young, so we recommend a more extensive research including a broader range of participants. Also, there are limitations related to the driving simulator, where the external validation is an often-mentioned issue even though the method has many advantages. In other words, a fixed-base simulator used in the research does not provide a completely realistic, real-life driving feeling. However, this disadvantage has been somewhat reduced by including the sound and by conducting the research in a completely dark environment (in order to get a more realistic feeling of night-time conditions).

Acknowledgements The research was part of the project entitled "Analitički postupci utvrđivanja utjecaja čimbenika cesta, čovjek, vozilo na događanje prometnih nesreća" financed by the Ministry of the Interior of the Republic of Croatia.

References

Anstey KJ, et al (2012) The role of cognitive and visual abilities as predictors in the multifactorial model of driving safety. Accid Anal Prev 45:766–74. https://doi.org/10.1016/j.aap.2011.10.006

Ariën C, et al (2017) The effect of pavement markings on driving behaviour in curves: a simulator study. Ergonomics 60(5):701–713. https://doi.org/10.1080/00140139.2016.1200749

Babić D, Babić D, Ščukanec A (2017) The impact of road familiarity on the perception of traffic signs—eye tracking case study. In: The proceedings of the 10th international conference: "Environmental Engineering", Vilnius, Lithuania, pp 1–7

Bener A, Crundall D (2008) Role of gender and driver behaviour in road traffic crashes. Int J Crashworthiness 13(3):331–336. https://doi.org/10.1080/13588260801942684

Biancardo SA, et al (2019) Design criteria for improving safety performance of rural intersections. J Adv Trans Article 1232058. https://doi.org/10.1155/2019/1232058

Burns DM, Johnson NL (1997) Characterizing the visual performance of fluorescent retroreflective signing materials using the fluorescent luminance factor. In: Proceedings of the 8th congress of the international colour association, vol 1, pp 359–362

European Commission (2018a) Road. Available at https://ec.europa.eu/transport/road_safety/sites/roadsafety/files/pdf/ersosynthesis2018-roads.pdf

European Commission (2018b) Traffic safety basic facts: junctions. Available at https://ec.europa.eu/transport/road_safety/sites/roadsafety/files/pdf/statistics/dacota/bfs2018_junctions.pdf

European Commission (2019) 2018 road safety statistics: what is behind the figures? Available at https://europa.eu/rapid/press-release_MEMO-19-1990_en

Federal Highway Administration (2019) Intersection safety. Available at https://safety.fhwa.dot.gov/intersection/

Ferko M, et al (2019) The impact of traffic sign quality on the frequency of traffic accidents. Promet Traffic Trans 31(5):549–558. https://doi.org/10.7307/ptt.v31i5.3023

Gray RC, Quddus MA, Evans A (2008) Injury severity analysis of accidents involving young male drivers in Great Britain. J Saf Res 39:483–495. https://doi.org/10.1016/j.jsr.2008.07.003

Gregersen NP, Bjurulf P (1996) Young novice drivers: towards a model of their accident involvement. Accid Anal Prev 28(2):229–241. https://doi.org/10.1016/0001-4575(95)00063-1

Hallmark SL, Hawkins N, Smadi O (2012) Evaluation of low-cost treatments on rural two-lane curves. Report No. IHRB Project TR-579. Midwest Transportation Consortium, Iowa Department of Transportation and Iowa Highway Research Board, Ames, IA

Himes S, Porter RJ, Eccles K (2018) Safety evaluation of geometric design criteria: intersection sight distance at unsignalized intersections. J Trans Res Board 2672(39):11–19. https://doi.org/10.1177/0361198118783162

Hummer JE et al (2010) Curve crashes: road and collision characteristics and countermeasures. J Trans Saf Secur 2(3):203–220. https://doi.org/10.1080/19439961003734880

Hussain AH, et al (2019) Impact of perceptual countermeasures on driving behavior at curves using driving simulator. Traffic Inj Prev 20(1):93–99. https://doi.org/10.1080/15389588.2018.1532568

Li Z et al (2018) Exploring driver injury severity patterns and causes in low visibility related single-vehicle crashes using a finite mixture random parameters model. Anal Method Accid Res 20:1–14. https://doi.org/10.1016/j.amar.2018.08.001

Mackenzie AK, Harris JM (2017) A link between attentional function, effective eye movements, and driving ability. J Exp Psychol Hum Percept Perform 43(2):381–394. https://doi.org/10.1037/xhp0000297

McGee HW, Hanscom FR (2006) Low-cost treatments for horizontal curve safety. Report No. FHWA-SA-07-002. Federal Highway Administration, Washington, DC

Montella A, et al (2015) Effects of traffic control devices on rural curves driving behaviour. J Trans Res Board 2492(1):10–22. http://dx.doi.org/10.3141/2492-02

Plainis S, Murray IJ, Pallikaris IG (2006) Road traffic casualties: understanding the night-time death toll. Inj Prev 12(2):125–128. https://doi.org/10.1136/ip.2005.011056

Polus A (1985) Driver behaviour and accident records at unsignalized urban intersections. Accid Anal Prev 17(1):25–32. https://doi.org/10.1016/0001-4575(85)90005-3

Priambodo MI, Siregar1 ML (2018) Road sign detection distance and reading distance at an uncontrolled intersection. Int Conf Civil Environ Eng 65:1–8. https://doi.org/10.1051/e3sconf/201865 09004

Rice TM, Peek-Asa C, Kraus JF (2009) Nighttime driving, passenger transport, and injury crash rates of young drivers. Inj Prev 9(3):245–250. https://doi.org/10.1136/ip.9.3.245

Šarić Ž, et al (2018) Identifying the safety factors over traffic signs in state roads using a panel quantile regression approach. Traffic Inj Prev 19(6):607–614. https://doi.org/10.1080/15389588. 2018.1476688

Schnell T et al (2001) Legibility distances of fluorescent traffic signs and their normal color counterparts. J Trans Res Board 1754:31–41. https://doi.org/10.3141/1754-04

Scott-Parker B, Oviedo-Trespalacios O (2017) Young driver risky behaviour and predictors of crash risk in Autralia, New Zealand and Colombia: Same but different? Accid Anal Prev 99:30–38. https://doi.org/10.1016/j.aap.2016.11.001

Sullivan JM, Flannagan MJ (2002) The role of ambient light level in fatal crashes: inferences from daylight saving time transitions. Accid Anal Prev 34(4):487–498. https://doi.org/10.1016/S0001-4575(01)00046-X

Thurston P (2009) Pavement markings role in enhancing road safety strategies. Roadmarking Industry Association of Australia, Australia

van Winsum W (2018) The effects of cognitive and visual workload on peripheral detection in the detection response task. Hum Factors 60(6):855–869. https://doi.org/10.1177/001872081877 6880

van Winsum W (2019a) A threshold model for stimulus detection in the peripheral detection task. Trans Res Part F 65:485–502. https://doi.org/10.1016/j.trf.2019.08.014

van Winsum W (2019b) Optic flow and tunnel vision in the detection response task. Hum Factors 61(6):992–1003. https://doi.org/10.1177/0018720818825008

World Health Organization (2018). Available at https://www.who.int/news-room/fact-sheets/detail/the-top-10-causes-of-death

World Road Association (2003) PIARC technical committee on road safety. Road safety manual

Xu X, et al. (2018) Accident severity levels and traffic signs interactions in state roads: a seemingly unrelated regression model in unbalanced panel data approach. Accid Anal Prev 120:122–129 (2018). https://doi.org/10.1016/j.aap.2018.07.037

Yang S et al (2019) What contributes to driving behavior prediction at unsignalized intersections? Trans Res Part C Emerg Technol 108:100–114. https://doi.org/10.1016/j.trc.2019.09.010

Yanko MR, Spalek TM (2013) Route familiarity breeds inattention: a driving simulator study. Accid Anal Prev 57:80–86. https://doi.org/10.1016/j.aap.2013.04.003

Zwahlen HT, Schnell T (1977) Superior traffic sign, pedestrian, bicycle, and construction worker conspicuity through the use of retroreflective fluorescent color materials. In: The proceedings of the 13th triennial congress of the international ergonomics association, vol 6, pp 484–486

Traffic Flow Simulators with Connected and Autonomous Vehicles: A Short Review

Filip Vrbanić, Dino Čakija, Krešimir Kušić, and Edouard Ivanjko

Abstract Autonomous Vehicles (AVs) and Connected Autonomous Vehicles (CAVs) are being widely tested and rapidly developed over the past few years. With the development and increasing number of AVs and CAVs in mixed traffic flow, it is necessary to analyze their impact on traffic safety, flow, speed, fuel consumption, and emissions. Also, appropriate traffic control algorithms need to be developed before they can be fully implemented and integrated into the traffic environment. To do so, such mixed traffic flows must be simulated in various traffic scenarios. Traffic flow simulators paired with communication network simulators are commonly used to perform multiple simulations of such traffic flows. In this paper, three often used traffic simulators VISSIM, AIMSUN, and SUMO paired with network simulators NS-3 and OMNET++ with their features to model AVs and CAVs in a simulation environment are analyzed. According to currently available and tested simulators in the research community, the most used ones were compared. Results of the synthesized technical aspects of each suggest that AIMSUN Next is more suitable for a less complex traffic model. At the same time, VISSIM is more suitable for a more complex traffic model.

Keywords Autonomous vehicle · Connected autonomous vehicle · Mixed traffic flow · Traffic simulator · Communication network simulator

F. Vrbanić (✉) · D. Čakija · K. Kušić · E. Ivanjko
Faculty of Transport and Traffic Sciences, University of Zagreb, Zagreb, Croatia
e-mail: filip.vrbanic@fpz.unizg.hr

D. Čakija
e-mail: dino.cakija@fpz.unizg.hr

K. Kušić
e-mail: kresimir.kusic@fpz.unizg.hr

E. Ivanjko
e-mail: edouard.ivanjko@fpz.unizg.hr

M. Petrović and L. Novačko (eds.), *Transformation of Transportation*, EcoProduction,
https://doi.org/10.1007/978-3-030-66464-0_2

1 Introduction

What society strives for is a safe, low cost, efficient, and environmental-friendly mobility. With the rapid development of Autonomous Vehicles (AVs) and Connected Autonomous Vehicles (CAVs), new possibilities and opportunities arise. Even though it is not precisely known when the deployment of AVs will be, the share of AVs in the consumer markets is projected to grow to over 70% by the year 2050 (Lavasani et al. 2016). Due to the increase of AV development in the United States of America, the National Highway Traffic Safety Administration (NHTSA) released the Federal Policy on Automated Vehicles in 2018 (NHTSA 2017). It includes guidelines for AV manufacturing and regulation. United States Department of Transportation also issued guidelines for AV development (US DOT 2020), and the Society of Automotive Engineers (SAE) adapted the levels of automation of road vehicles that range from level 0 (no automation) to level 5 (full automation) (SAE International 2018). With a substantial increase of knowledge and continuous improvements in space perception and computational power, autonomous driving features of some level are being introduced in almost every new vehicle. Adaptive Cruise Control (ACC), lane control, and automatic emergency braking have already been introduced by all major brands signifying the research efforts being put into self-driving cars.

Vehicle connectivity has also been significantly developed in recent years. Connectivity technology has enabled improved safety, performance, and cooperation among vehicles. The concept of platooning was demonstrated in the California PATH program (Shladover 2007). This concept denotes a group of vehicles traveling at small headway distance to increase the throughput on highways with the use of Vehicle-to-Vehicle (V2V) communication to coordinate multivehicle maneuvers and to exchange vehicle states that enable short headway time between the platooned vehicles. CAVs enable a variety of applications in Intelligent Transportation Systems (ITS), including traffic control, cooperative driving, improved safety, and more efficient energy consumption (Knight 2015). Although the power consumption of the sensors, computing, and communication equipment of CAVs needs to be taken into account (Gawron et al. 2018).

Deployment of CAVs and AVs is still in its early stages. The testing of prototypes in the real traffic environment, according to legal aspects, is not defined yet. There is still a gap and insufficient traffic data about the influence and interaction of AVs and CAVs on traffic flow containing human drivers. Since the penetration rate of such vehicles in real traffic is low, to get insight into their interaction with human drivers, various microscopic simulators are used to mimic the AVs and CAVs within the mixed traffic flows. Microscopic traffic simulators are mostly used to simulate each individual vehicle. The most commonly used commercial microscopic traffic simulators that have options for CAVs and AVs are VISSIM and AIMSUN, and the most widely used open-source simulator is SUMO. To model connectivity and communication between vehicles, network simulators such as NS-3 and OMNET++ are commonly used. In this paper, we analyze the possibilities of the mentioned traffic simulators to model and simulate AV and CAV behavior as close as possible to real-world scenarios and environments from the traffic simulator perspective. The analysis showed that all of them have features for CAV and AV implementation that can be extended and modeled via their respective interfaces, although AIMSUN Next is a complete simulator for CAVs with its built-in features.

2 Mixed Traffic Flows

The term mixed traffic flows refers to traffic flows that contain regular human-driven vehicles, AVs, and CAVs with different penetration rates. AVs represent vehicles that can get traffic information from different sensing technologies such as cameras, sensors, radars, and lidars integrated within the vehicle. They are characterized by high traffic law obedience, shorter time and space headway, and smaller gap acceptance. CAVs are similar to AVs with added features that allow communication to other vehicles (V2V), roadside infrastructure (Vehicle-to-Roadside—V2R) such as traffic signal controllers, infrastructure (Vehicle-to-Infrastructure communication—V2I) and the entire surrounding environment (Vehicle-to-Everything communication—V2X).

2.1 Characteristics of Autonomous Vehicles

The impact and development of AVs (also known as self-driving cars, driverless cars, or robotic cars) have been widely discussed due to significant improvements in Advanced Driver Assistance Systems (ADAS) and vehicular communications. AV is a vehicle that operates without human control and requires only a few or zero human interventions depending on its level of automation. In Campbell et al. (2010), it is stated that intra-vehicle communication, also known as Vehicle-to-Sensors (V2S) communication, enables AVs to sense their surrounding environment, detect and classify different objects, process all sensory information, and identify appropriate navigation paths while obeying traffic laws and regulations. The term autonomous driving refers to the control of vehicle motion in both the longitudinal and lateral direction. Autonomous driving systems include a variety of sensors for environment perception that include lidars, radars, cameras, and ultrasonic sensors, which enable autonomous driving in stochastic traffic environments. In urban areas, where GPS accuracy is low, a high-resolution perception system, combined with high-resolution environment maps, can provide precise positioning for the autonomous driving system. It is essential to combine a variety of technologies from different disciplines that include computer science, mechanical engineering, electronics engineering, electrical engineering, control theory, etc. to carry out successful autonomous navigation in such situations (Deshpande 2014). In 2018, SAE International released a revised document for levels of driving automation that defines six different levels (SAE International 2018). Table 1 shows the automation level description and driver support, including

Table 1 SAE level definition of driving automation

Level	Description	Required human behavior	Features
Level 0	Driver support features with no driving automation	Human is driving a vehicle even if support features are engaged	Limited to providing warnings and momentary assistance
Level 1	Driver support features with no driving automation	Human is driving a vehicle even if support features are engaged	Steering or brake/acceleration support
Level 2	Driver support features with no driving automation	Human is driving a vehicle even if support features are engaged	Steering and brake/acceleration support
Level 3	Automated driving features that drive the vehicle under limited conditions	When the feature requests, a human must drive the vehicle	Traffic jam chauffeur
Level 4	Automated driving features that drive the vehicle under less limited conditions	Does not require human to take over driving	Driverless taxi, the steering wheel may not be installed
Level 5	Automated driving features that drive the vehicle under any conditions	Does not require human to take over driving	Fully autonomous driving under any conditions

automated features. Level 0 requires full human control. Features provided to help the driver include systems like automatic emergency braking, blind-spot warning, and lane departure warning. Level 1 still requires full human engagement to operate the vehicle. Features provided here include lane-centering or ACC. Level 2 is similar to level 1, but the mentioned features can be used simultaneously. In level 3, most of the driving tasks are operated by the control systems for automated driving but may request a human to operate the vehicle in uncertain situations. Level 4 represents full autonomous driving mode, and such a vehicle may not even include a steering wheel. If some uncertainty occurs in automated vehicle systems, the vehicle can abort the operation and safely stop driving. Level 5 refers to a fully autonomous vehicle without any human intervention.

2.2 *Characteristics of Connected and Autonomous Vehicles*

CAV is a technology that enables a connectivity feature that potentially reduces the number of traffic incidents and improves the efficiency of traffic in urban environments. CAV is defined as a vehicle that can operate by automated driving and sharing information by connectivity with other vehicles, traffic participants, the road infrastructure, and the cloud (V2X). CAVs showcased the positive effects on urban traffic in comparison to present experiences with higher throughput with the reduction of distance between vehicles that results in increased capacity by not raising delay times (Bajpai 2016). It is also noted that the number of owned vehicles could decrease by sharing a vehicle through third-party companies or individuals. Thus, the number of vehicles per household may drop with the availability of CAVs. Such a vehicle can return home after mobility service is performed to be used by other household members. In Cagney (2017), it is noted that people who do not have a driving license would benefit significantly from CAVs by reducing the use of public transport by using CAVs for transportation instead. As mentioned, traffic incidents will also potentially be decreased by eliminating human error in AVs and CAVs. It is estimated that the occurrences of traffic accidents will be decreased by 50% by the year 2040 using these technologies (Cagney 2017).

Vehicle connectivity and communication is a critical requirement in CAV operation in the domain of ITS. Communication is a crucial component to achieve multivehicle cooperation and awareness of obstacles outside the line of sight of one vehicle. Technologies used to achieve such communication are Dedicated Short Range Communications (DSRC) and cellular communication (4G and 5G). DSRC can be used as a one or two-way short-range to midrange wireless channel that enables the transmission of very high data transmission rates. There are two types of DSRC that include V2V, V2R, and V2I. Both types need a protected wireless interface constancy in short time delays, even in extreme weather conditions. Applications of DSRC include safety warnings, intersection assistance, traffic condition broadcasting, and payment of tolls and parking. The advantages of DSRC technology include security, low latency, interoperability, and resilience to extreme weather conditions. However, it requires dedicated hardware. On the other hand, 5G may also compete with DSRC for V2V and V2I communications since it offers access to cloud services with lower latency and cooperation with other vehicles and infrastructure. V2R connectivity is also a way of communication that uses DSRC technology to enable connections between vehicles and infrastructure, such as traffic signal controllers, street signs, and roadside sensors. Figure 1 shows the possible use cases of DSRC in a V2V communication environment (Kenney 2011).

V2I enables connection to the Internet so vehicles can share information via a roadside base station. It captures and transmits data such as traffic congestion and weather conditions, and wirelessly transmits processed data. Smart traffic signals powered by V2I help drivers to percept the traffic conditions better and to accurately estimate arrival times, which can improve communication between delivery drivers and customers. V2I aids in better driver-assistance systems that enable CAVs, which could aid in future urban infrastructure planning.

V2X includes both V2V and V2I communication technology that allows vehicles to communicate with the traffic environment, including other vehicles and infrastructure. V2X technology allows other traffic entities to notify drivers of dangerous weather conditions, nearby incidents and traffic congestion, and other close-range hazardous behaviors. V2X provides the sharing of valuable information directly to the vehicle, reducing the reaction time it takes for the driver to respond. Similar to V2I and V2V technology, V2X will be most effective when every traffic entity is equipped with this connected vehicle technology. Figure 2 shows an overview of vehicle communication technology (Lu et al. 2014). In Zhou et al. (2020), the authors analyzed traditional V2X technologies that are evolving to the Internet of

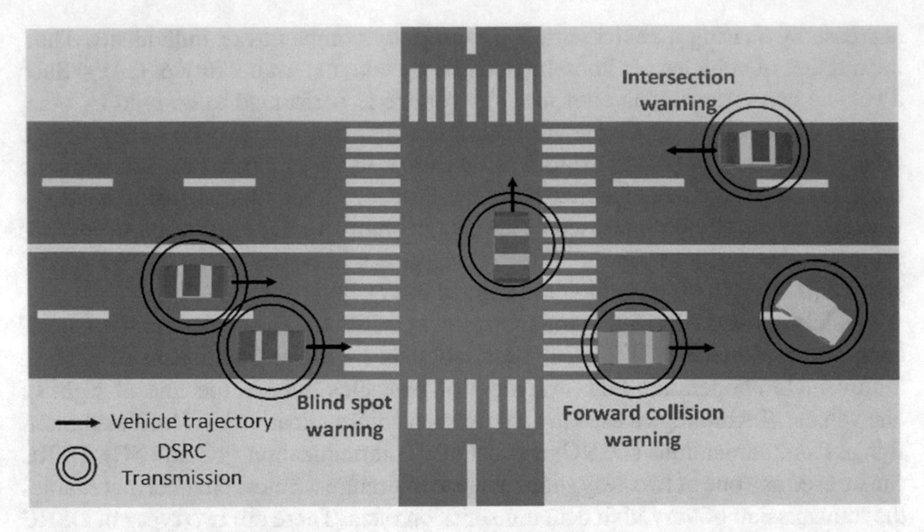

Fig. 1 Vehicles sending safety messages, displaying in-vehicle warnings in DSRC V2V environment

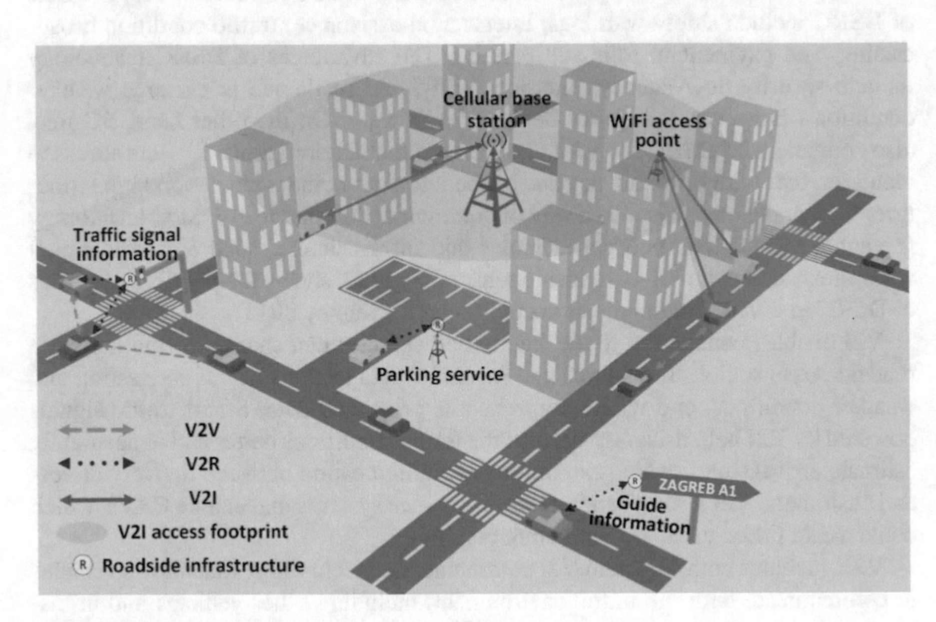

Fig. 2 Vehicle communication technologies—overview

Vehicles (IoV) for covering the increasing demands of emerging advanced vehicular applications, such as ITS and AVs. It is noted that in recent years, IoV technologies have been developed and achieved significant progress. Survey (Zhou et al. 2020) on evolutionary V2X technologies toward IoV provides insights into IoV development.

2.3 *Microscopic Vehicle Simulation Models*

Simulation models for CAVs require adjusted car-following models, lane changing models, conflict resolution, and communication models. Conventional car-following models are developed based on human driving behavior, which defines interactions with the preceding vehicle. However, CAVs use a different car-following model and characteristics. Built-in sensing and communication technology allow CAVs to comprehend the down-stream traffic situations beyond human drivers' viewpoint. With the use of Cooperative Adaptive Cruise Control (CACC), CAVs can communicate (using V2V) with each other to influence traffic flow characteristics. Recently, research efforts have been active to develop intelligent-vehicle car-following models by enhancing the conventional car-following model. Treiber et al. (2000) proposed Intelligent Driver Model (IDM) to simulate occurring bottleneck congestions. IDM provides collision-free behavior and self-organized characteristic that differ from traditional microscopic models. It also uses ACC as it allows for smoother traffic flow behavior with fewer speed oscillations and no sudden decelerations of the preceding vehicle. Cooperative Intelligent Driver Model (CIDM) is used to simulate the driving behaviors of AVs with a cooperative driving strategy (Zhou et al. 2016). The second part of the CAV models is the lane changing model. It includes a vehicle merging model, which, as a mandatory type of lane change, can significantly influence mainstream traffic flow. CAVs on the main road can adjust their acceleration and form a longer headway gap in advance. Thus, the entering vehicles can safely merge into the mainstream flow.

CAVs also must have the capability of sending and receiving information to and from other vehicles and infrastructure with the assumption that reliable connectivity is achieved. However, this information may always not be available, and locations and drivers' behavior may change according to the amount of information they receive. With active V2V, drivers are certain about the behavior of other vehicles. They are also aware of the driving environment, road condition, and weather conditions downstream of their current location. Connectivity in V2V and V2I networks in a simulation framework is needed to achieve and enable the flow of information that is necessary for a stable connected driving environment. Therefore, it is necessary to simulate wireless communications. Commonly used simulators for V2V/V2I communication networks are NS-3 and OMNET++ that can be combined with traffic simulators ("NS-3 network simulator" n.d.; "OMNET++ Discrete Event Simulator" n.d.).

3 Analyzed Microscopic Traffic Simulators

A microscopic traffic simulation is an important tool for any general traffic analysis. They are used to analyze traffic through a detailed representation of the behaviors of individual vehicles. It is important to formulate microscopic models that capture the capabilities of the new technologies as well as the vehicle behavior to explore the impacts of AVs and CAVs on traffic flows. For human drivers, one could rely on a variety of existing models. Specific logic for autonomous vehicles is robotic and necessarily supplied by the operating entity. Connected vehicle behavior is mainly dependent on the implemented capabilities. For a specific purpose, vehicle types can be modeled and tested in a developed environment with the use of traffic simulators.

3.1 PTV VISSIM

PTV VISSIM is a commercial microscopic traffic simulator that allows simulation of individual vehicle movement. It supports built-in pedestrian and vehicle simulation in a single software platform. It allows realistic localized conditions to be replicated with a detailed representation of road geometry and vehicle behavior models. It gives a realistic and detailed overview of the traffic flow and impacts, with the possibilities to define multiple what-if scenarios. VISSIM internal functions allow adaptation of car following, lane change, and vehicle speed parameters models. A simple method to model some of AVs related features within a given traffic network include smaller standstill distances, smaller lateral distances, acceleration rates from a standstill, creating space for merging vehicles, and many more. Possible applications of internal CAV functions allow investigation of the influence of different follow-up distances or acceleration oscillations on traffic flow at constant or varying AV penetration rates. VISSIM external functions allow external modeling of AVs and CAVs with the utilization of several interfaces. Thus, the control logic of the AVs can be defined externally. Three different interfaces are available in VISSIM: (i) DriverModel.DLL interface, (ii) DrivingSimulator.DLL interface, and (iii) COM Interface (PTV VISSIM 2017a, b). First, the DriverModel.DLL interface is suitable for simulation of AVs interacting with other vehicles. Second, the DrivingSimulator.DLL interface allows testing of the interactions between AVs (simulated by other software), and vehicles and pedestrians. Third, the COM Interface is suitable for the development of algorithms to control the AV behavior and other processes like traffic lights.

3.2 SUMO

SUMO is an open-source microscopic traffic simulator (Behrisch et al. 2011). Each vehicle is defined by an identifier (ID), the departure time, and the vehicle's route through the network, but can also be described with more details. The departure and arrival properties, vehicle type (class) assigned, which represents the vehicle's physical properties, and the variables of the used movement model can be modeled (Behrisch et al. 2011). For large-scale scenarios, Origin/Destination (O/D) matrices are used to define trips between traffic assignment zones (Behrisch et al. 2011).

SUMO software includes the "od2trips" command for converting O/D matrices to single-vehicle trips that assign vehicles to edges of the road network as departing and arrival position (Behrisch et al. 2011). Routes are calculated by performing a traffic assignment employing a routing procedure such as shortest path calculations under different cost functions (Sagir and Ukkusuri 2018). SUMO provides Traffic Control Interface (TraCI) that provides statistical data and direct access to elements (Wegener et al. 2008). SUMO does not provide AV simulation directly. It can be done with integration with traffic and robotic simulator. A bidirectional communication should take place, with the AV simulator providing kinematic variables to the traffic simulator that then calculates its surroundings and return its data to the AV simulator to integrate the two simulators in time and space (Pereira and Rossetti 2012). For an integrated CAV simulation environment, the use of frameworks that integrate SUMO with a communication network simulator is used. The coupling is done via SUMO's TraCI-API that allows arbitrary runtime influence on the simulation behavior. The vehicle movements can be modeled either with lane-level accuracy (vehicles always in the center of their lanes) or with sublane accuracy if required. A range of models can be configured, or adjusted models can be added in the simulator to model the behavior of particular AVs. Authors in Wu et al. (2019) developed an extension for SUMO called FLOW. It is a modular framework for machine learning that eases the configuration and composition of modules, to enable learning control laws for AVs in complex traffic settings involving nonlinear vehicle dynamics and arbitrary network configurations. The implementation was built upon SUMO for vehicle and traffic modeling, Ray RLlib library in Python for reinforcement learning, and OpenAI gym for the Markov Decision Process.

3.3 *AIMSUN Next*

AIMSUN Next simulator allows the modeling of small and large transportation networks. Python scripting allows automated operations such as modifying model parameters, importing and exporting data, performing additional calculations on simulation outputs, and launching multiple simulations or assignments. AIMSUN provides an optional V2X Software Development Kit (V2X SDK) feature that is intended to include Vehicular Ad hoc Network (VANET) communications in a microscopic simulation-based on industry-standard protocols. The V2X SDK also includes components required to implement a "Rules Engine" in the vehicle, roadside unit, and traffic management center. The External Agent Interface (EAI) is designed to introduce externally controlled vehicles to a microscopic simulation and guided by the actions of external agents such as a human driver in a simulator, or an AV controller, or by an experimental control system being tested in the simulation environment ("Aimsun Next" n. d.). AIMSUN Next has features for modeling CAVs that include adjusting car following, lane changing, and gap acceptance behavioral models. CACC model and merging behavior can also be modified by adjusting cooperation and aggressiveness parameters.

3.4 Communication Network Simulators

Mostly used network simulators are NS-3 and OMNeT++. NS-3 is an open-source discrete-event network simulator for Internet systems used for research and education ("NS-3 network simulator" n.d.). It supports a large simulation scenario with up to 20,000 nodes for a more realistic simulation. In Liu et al. (2016), the adapted client/server communication model was used for communication with NS-3 and SUMO via the TraCI interface. NS-3 uses 'while loop' to control SUMO's time frame. NS-3 requests the vehicle's position in a SUMO scenario every millisecond and updates it in NS-3. After running the simulation protocol, NS-3 dynamically changes the vehicle's driving route in the SUMO scenario through communication.

OMNeT++ is an extensible, modular simulation library and framework for building network simulators. It includes wired and wireless communication networks, on-chip networks, queueing networks, and more ("OMNET++ Discrete Event Simulator" n.d.). Authors of (Sommer and Dressler 2010) developed an open-source framework Veins for running vehicular network simulations. It is based on two simulators: OMNeT++ and SUMO that offers a comprehensive suite of models for Inter-Vehicle Communication (IVC) simulation.

4 Discussion

The main problem for simulating (C) AVs is the lack of driver behavior models, which are different compared to human drivers' models like the Wiedemann, IDM, or Gipps models that are widely used today (Čakija et al. 2019). VISSIM has a built-in car-following model that is a psychophysical model developed by Wiedemann (Shen et al. 2018), specifically Wiedemann74 for urban and Wiedemann99 for freeway scenarios. The downside is that there is a limited number of (C) AVs reactions to any given situation, and the simulation of V2I is not possible. In Shen et al. (2018), a car following rate control algorithm based on the Wiedemann model for VANET was presented. Authors in Virdi et al. (2019) provide insights about the safety improvements by incrementally transitioning the fleet to CAVs using custom-developed Virdi CAV Control Protocol (VCCP) and calibrated microsimulation environment in VISSIM (Virdi et al. 2019). Their simulation results showed that the number of accidents involving human-driven vehicles decreases as the rate of CAVs increases. In Khondaker and Kattan (2018), the authors present a Variable Speed Limit (VSL) control algorithm that simultaneously maximizes the mobility, safety, and environmental benefit in CAVs environment. This approach was evaluated by VISSIM's microsimulation tool via an integrated COM—Matlab interface and VSL control logic was developed using Visual Basic for Applications (VBA) (Khondaker and Kattan 2018).

SUMO was used in Rakkesh et al. (2017) for managing vehicular traffic flow while platoon management and vehicular communications were jointly handled using VEINS, OMNET++, and Platooning Extension for Veins (PLEXE) developed by authors in Segata et al. (2014). PLEXE framework was used to schedule an 8-node platoon of vehicles to assess the efficiency of platooning by creating different scenarios with closer and higher inter-vehicular gaps using CACC and ACC (Rakkesh et al. 2017). In Fernandes and Nunes (2010), SUMO was used for multi-vehicles studies in urban scenarios. SUMO was connected with other applications through a client-server architecture with TraCI to control vehicle parameters on run time, such as platoon leaders that allow vehicle cooperation and coordination. At the same time, the SUMO acts as the simulation engine for the platoon control. The platoon leaders' parameters were controlled externally with the TraCI package enabling platoons' control through their leaders, and the SUMO itself controlling the remaining vehicles through a module that implements the longitudinal control model of the vehicles (Fernandes and Nunes 2010). Autonomous robo-taxis were analyzed in the city of Milan using the SUMO simulation framework and enhancing the surrounding functionality (Alazzawi et al. 2018). Simulation results show reduced congestion and vehicle emissions. The authors extended the existing SUMO functionality by enhancing its framework by accessing the simulations at each simulation step via the TraCI interface to analyze prospects of shared and autonomous robo-taxis. In the recent study (Lin et al. 2014), a protocol called Cooperative Driving Control Protocol (CDCP) for intelligent AVs that defines a standard with common messages and format for AVs was developed. The typical standard format and definitions of the CDCP packet make the AVs have a common language to achieve cooperated intelligent driving with other vehicles, decrease the reaction time, and decrease the average travel time at the same speed (Lin et al. 2014). SUMO and network simulator NS-3 were used to simulate wireless transmission and intelligent autonomous vehicle mobility. Each AV was equipped with an on-board unit that contains a VANET communication interface with a transmission range of 250 meters. Thus, AVs can communicate with neighboring vehicles. Implementation of the lane-merging maneuver in the SUMO simulator was proposed as it may be implemented in real-world scenarios in the future (Domingues et al. 2018). Two different mechanisms for platoon formation were discussed: a take-or-leave proposal and a naive-based negotiation. The simulation results show that platooning has an impact on reducing fuel consumption and CO2 emissions. The effect of lane-merging techniques and negotiation with platooning is highlighted through the significant time loss incurred in the scenario where no lane-merging methods were applied to the simulation (Domingues et al. 2018).

The recent study (Dandl et al. 2017) analyzed autonomous aTaxi using a calibrated AIMSUN model of the city of Munich. An autonomous taxi system was implemented into an existing, calibrated traffic model. The impacts of an aTaxi system on the traffic network for selected scenarios in the city of Munich was tested (Dandl et al. 2017). Results show that even in the scenario, where 10% of the private vehicle trips originating and ending in the study area were exclusively replaced by aTaxi requests an extra network-wide load increase of private vehicle delay times by 1% only due to empty rides. An AIMSUN traffic simulation model was also used in Mattas et al. (2018) where three types of vehicles were simulated to construct a mixed traffic flow containing manually driven vehicles, AVs without human control, and CAVs that are operating as AVs with the use of connectivity. For the simulation of manually driven vehicles, the default modified Gipps' car-following model was used. A first-order model representing ACC vehicle longitudinal behavior was used to simulate AVs, which do not have the connectivity features. The default AIMSUN model was used for the lateral movement, according to the ACC maximum deceleration and car following deceleration. AVs and CAVs were forced to obey the speed limits, in contrast to manually driven vehicles that have an acceptance factor allowing them to accede to the speed limit. For CAVs that do not exchange information with nearby vehicles, AV model was used when following another type of vehicle, and lane changing was modeled based on the default AIMSUN algorithm, using the CAVs particular car following deceleration model (Mattas et al. 2018). The authors used the microscopic traffic simulator AIMSUN with its microSDK tool to overwrite the default vehicle behavioral models with the IDM car-following model (Perraki et al. 2018). The default Gipps lane-changing model was complemented at merging network locations with some heuristic rules that consist of a set of inequality conditions considering the vehicles current state, the neighboring traffic conditions and three variables as current speed, the relative speed to the target-lane vehicles, and the available gap in the target lane (Perraki et al. 2018). The heuristic lane changing rules are applied at the on-ramp and lane-drop areas, while in the rest of the motorway model, the Gipps lane-changing model was used.

Table 2 shows a brief overview of the analyzed traffic simulators. All of them have their respective advantages and disadvantages. Provided features are extendable and configurable via different interfaces, as mentioned in Sect. 3. Analyzed simulators

Table 2 Traffic simulators overview

	VISSIM	AIMSUN	SUMO
Pricing	Commercial. Dependent on the version and modules	Commercial. Dependent on the version and mo-dules	Open-source
Expandability	COM interface, DriverModel.DLL, and DrivingSimulator.DLL interfaces	EAI, Adaptive control interface, GIS and CAD interface, Planning software interfaces	TraCI interface
Vehicle models and options	Very wide variety	Wide variety	Wide variety
Advantages	Versatility, flexible API	Easy to use, very flexible API	Strong community, good manuals
Disadvantages	Lack of vehicle communication module	Fewer model-specific options	Lack of vehicle communication module, merging behavior must be adjusted
Complexity	Less complex	Less complex	Complex

can be integrated and supplemented with both robotic and network simulators via the use of their respective interfaces (PTV VISSIM 2017a, b; Wegener et al. 2008; "Aimsun Next" n. d.). VISSIM and AIMSUN are both very powerful commercial simulators, but AIMSUN Next provides the best user experience, functionality, and extensibility, and is suitable to use for a less complex traffic model since it requires more processing power than VISSIM. For more extensible and complex traffic model, VISSIM is more appropriate, since it offers a wide variety of vehicle models and parameters' adjustment (PTV VISSIM 2017a, b).

5 Conclusion

Vehicle automation and connectivity are being developed extensively in recent years. The role of traffic control and planning is crucial for the improvement of the safety and performance of the existing road networks. This paper surveyed the current literature on microscopic simulators to test and simulate AV and CAV impact on traffic in mixed traffic flows. The main components and technologies for CAVs are discussed. Traffic simulators alongside with robotic and network simulators make a simulation framework for testing of mixed traffic flows in appropriate scenarios. It is essential to analyze the impact CAVs, and AVs may have on traffic flows, safety, vehicle emissions, and urban mobility. While testing in real-world scenarios is still very challenging, the current state of technology offers the opportunity to deploy those vehicles. However, it is essential to simulate their impact before real-world implementation. Selecting appropriate testing scenarios, which are representative of real-world traffic conditions, is a non-trivial open question. The current direction pursued by authorities, manufacturing companies, and academic researchers support this view.

The connectivity possibilities on which CAV technology is based on is also analyzed. The existing vehicle modeling approaches can be framed within two commercial microscopic traffic simulators; VISSIM and AIMSUN, and one open-source simulator SUMO. Most recent papers were presented, and key modeling features were pointed out regarding AVs and CAVs. Finally, the use of simulation frameworks implemented with traffic simulators based on real-world traffic data and realistic simulation scenarios is key for experimental validation and development of the beforementioned vehicles and required infrastructure. Analyzed simulators are usable for CAV and AV simulation. Although AIMSUN provides built-in vehicle communication module V2X SDK, VISSIM and SUMO require the use of network simulators to do so. Future work on simulators must be on including built-in and easy to use communication modules and more versatile vehicle models for CAVs.

Acknowledgements The authors thank the companies PTV Group for providing a VISSIM research license, and AIMSUN for providing an AIMSUN NEXT classroom license. This work has been partly supported by the University of Zagreb and Faculty of Transport and Traffic Sciences under the grants "Investigation of the impact of autonomous vehicles on urban traffic flow characteristics" and "Innovative models and control strategies for intelligent mobility", and by the European Regional Development Fund under the grant KK.01.1.1.01.0009 (DATACROSS). This research has also been carried out within the activities of the Centre of Research Excellence for Data Science and Cooperative Systems supported by the Ministry of Science and Education of the Republic of Croatia.

References

Aimsun Next (n.d.). Available at https://www.aimsun.com/aimsun-next/. Accessed 2 Feb 2020

Alazzawi S, Hummel M, Kordt P, Sickenberger T, Wieseotte C, Wohak O (2018) Simulating the impact of shared, autonomous vehicles on urban mobility - a case study of Milan. EPiC Series Eng 2:94–110

Bajpai JN (2016) Emerging vehicle technologies & the search for urban mobility solutions. Urban Plann Trans Res 4(1):83–100

Behrisch M, Bieker L, Erdmann J, Krajzewicz D (2011) SUMO—simulation of urban mobility an overview. Paper presented at the 3rd international conference on advances in system simulation, Barcelona, Spain, 23–29 October 2011

Cagney MR (2017) Autonomous vehicles: research report. http://mrcagney.com/case-studies/research/autonomous-vehicles-research-report/ Accessed 26 Jan 2020

Čakija D, Assirati L, Ivanjko E, Cunha AL (2019) Autonomous intersection management: a short review. Paper presented at the international symposium ELMAR, Zadar, Croatia, 23–25 September 2019

Campbell M, Egerstedt M, How J, Murray RM (2010) Autonomous driving in urban environments: approaches, lessons and challenges. Philos Trans Royal Soc A Math Phys Eng Sci 368(1928):4649–4672

Dandl F, Bracher B, Bogenberger K (2017) Microsimulation of an autonomous taxi-system in Munich. Paper presented at the 5th IEEE international conference on models and technologies for intelligent transportation systems (MT-ITS), Naples, Italy, 26–28 June 2017

Deshpande P (2014) Road safety and accident prevention in India: A review. Int J Adv Eng Technol 5(2):64–68

Domingues G, Cabral J, Mota J, Pontes P, Kokkinogenis Z, Rossetti RJF (2018) Traffic simulation of lane-merging of autonomous vehicles in the context of platooning. Paper presented at the 2018 IEEE international smart cities conference (ISC2), Kansas City, MO, USA, 16–19 September 2018

Fernandes P, Nunes UJ (2010) Platooning of autonomous vehicles with intervehicle communications in SUMO traffic simulator. Paper presented at the 13th international conference on intelligent transportation systems (ITSC 2010). Madeira Island, Portugal, 19–22 September 2010

Gawron JH, Keoleian GA, De Kleine R, Wallington TJ, Kim HC (2018) Life cycle assessment of connected and automated vehicles: sensing and computing subsystem and vehicle level effects. Environ Sci Technol 52

Kenney JB (2011) Dedicated short-range communications (DSRC) standards in the United States. Proc IEEE 99(7):1162–1182

Khondaker B, Kattan L (2018) Variable speed limit: a microscopic analysis in a connected vehicle environment. Transp Res Part C 58:146–159

Knight W (2015) Car-to-car communication. https://www.technologyreview.com/s/534981/car-to-car-communication/

Lavasani M, Jin X, Du Y (2016) Market penetration model for autonomous vehicles on the basis of earlier technology adoption experience. Trans Res Rec J Trans Res Board 2597:67–74

Lin C, Dong F, Hirota K (2014) A cooperative driving control protocol for cooperation intelligent autonomous vehicle using VANET technology. Paper presented at the 2014 joint 7th international conference on soft computing and intelligent systems (SCIS) and 15th international symposium on advanced intelligent systems (ISIS), Kitakyushu, Japan, 3–6 December 2014

Liu W, Wang X, Zhang W, Yang L, Peng C (2016) Coordinative simulation with SUMO and NS3 for vehicular Ad Hoc networks. Paper presented at the 22nd Asia-Pacific conference on communications, Yogyakarta, Indonesia, 25–27 Aug 2016

Lu N, Cheng N, Zhang N, Shen X, Mark JW (2014) Connected vehicles: solutions and challenges. IEEE Internet Things J 1(4):289–299

Mattas K, Makridis M, Hallac P, Alonso Raposo M, Thiel C, Toledo T, Ciuffo B (2018) Simulating deployment of connectivity and automation on the Antwerp ring road. IET Intel Transport Syst 12(9):1036–1044

NHTSA (2017) US DOT releases new automated driving systems guidance. https://www.nhtsa.gov/press-releases/us-dot-releases-new-automated-driving-systems-guidance. Accessed 21 Jan 2020

NS-3 network simulator (n.d.) Available at https://www.nsnam.org/. Accessed 3 Feb 2020

OMNET++ Discrete Event Simulator (n.d.) Available at https://omnetpp.org/. Accessed 3 Feb 2020

Pereira JLF, Rossetti RJF (2012) An integrated architecture for autonomous vehicles simulation. Artificial Intelligence and Computer Science Laboratory, Department of Informatics Engineering, Faculty of Engineering, University of Porto

Perraki G, Roncoli C, Papamichail I, Papageorgiou M (2018) Evaluation of a model predictive control framework for motorway traffic involving conventional and automated vehicles. Trans Res Part C 92:456–471

PTV Vissim (2017) PTV VISSIM & CONNECTED AUTONOMOUS VEHICLES. Available at http://www.sfbayite.org/wp-content/uploads/2017/04/1%20VISSIM_CAV_SFITE_April2017.pdf. Accessed 28 Jan 2020

PTV Vissim (2017) PTV VISSIM Webinar: Why Simulate Connected & Autonomous Vehicles on our Transport Systems?. Available at: https://www.ptvgroup.com/en/contact-support/webinars/. Accessed 28 Jan 2020

Rakkesh ST, Weerasinghe AR, Ranasinghe RAC (2017) An intelligent highway traffic model using cooperative vehicle platooning techniques. Paper presented at the 2017 Moratuwa engineering research conference, Moratuwa, Sri Lanka, 29–31 May 2017

SAE International (2018) Taxonomy and Definitions For Terms Related To Driving Automation Systems For On-Road Motor Vehicles J3016_201806. SAE international. https://saemobilus.sae.org/content/j3016_201806. Accessed 22 Jan 2020

Sagir F, Ukkusuri S (2018) Mobility impacts of autonomous vehicle systems. Paper presented at the 21st international conference on intelligent transportation systems, Maui, Hawaii, USA, 4–7 November 2018

Segata M, Joerer S, Bloessl B, Sommer C, Dressler F, Lo Cigno R (2014) PLEXE: a platooning extension for veins. Paper presented at the 2014 IEEE vehicular networking conference (VNC), Paderborn, Germany, 3–5 December 2014

Shen Z, Zhang X, Wang S, Yang D (2018) A car following based rate control algorithm for VANET in intelligent transportation systems. In: Proceedings of the 2018 IEEE wireless communications and networking conference, Barcelona, Spain, 15–18 April 2018

Shladover SE (2007) PATH at 20—history and major milestones. IEEE Trans Intell Trans Syst 8(4):584–592

Sommer C, Dressler F (2010) Bidirectionally coupled network and road traffic simulation for improved IVC analysis. IEEE Trans Mob Comput 10(1):3–15

Treiber M, Hennecke A, Helbing D (2000) Congested traffic states in empirical observations and microscopic simulations. Phys Rev E 62:1805–1824

US DOT (2020) Ensuring American leadership in automated vehicle technologies, automated vehicles 4.0, US DOT. https://www.transportation.gov/policy-initiatives/automated-vehicles/av-40. Accessed 3 Feb 2020

Virdi N, Grzybowska H, Waller ST, Dixit V (2019) A safety assessment of mixed fleets with connected and autonomous vehicles using the surrogate safety assessment module. Accid Anal Prev 131:95–111

Wegener A, Piorkowski M, Raya M, Hellbruck H, Fischer S, Hubaux JP (2008) TraCI: an interface for coupling road traffic and network simulators. Paper presented at the 11th communications and networking simulation symposium, April 2008

Wu C, Kreidlieh AR, Parvate K, Vitinsky E, Bayen AM (2019) Flow: a modular learning framework for autonomy in traffic. IEEE Transactions on robotics (in review). 2019 Available at https://arxiv.org/abs/1710.05465

Zhou M, Qu X, Jin S (2016) On the impact of cooperative autonomous vehicles in improving freeway merging: a modified intelligent driver model-based approach. IEEE Trans Intell Transp Syst 18(6):1422–1428

Zhou H, Xu W, Chen J, Wang W (2020) Evolutionary V2X technologies toward the internet of vehicles: challenges and opportunities. Proc IEEE 108(2):308–323

Application of Dimensionless Method to Estimate Traffic Delays at Stop-Controlled T-Intersections

Mohammad Ali Sahraei, Emre Kuşkapan, and Muhammed Yasin Çodur

Abstract Traffic delay is one of the essential aspects taken into consideration in the evaluation of operational performance of intersections. Delays at stop-controlled intersections are estimated using mathematical models that estimate the delays experienced by the minor road vehicles. However, the applicability of the existing mathematical models is subjected to the traffic local conditions. This paper discusses a current research carried out to develop a new model for the estimation of traffic delays experienced by the minor approach vehicles at stop-controlled T-intersections. Data pertaining to the analysis of traffic delays was collected at two stop-controlled intersections using a video camera recording technique. A dimensionless method was applied in the analysis of the data and two mathematical models for the estimation of delay to the right- and also left-turning manoeuvre on the minor approach were proposed. In this case, the comparisons between actual delays and the values predicted using the dimensionless method showed that there was no significant difference between the values. As such, it was suggested that the mathematical dimensionless-based models developed in this study are directly applicable, with a reasonable accuracy, for the analysis of traffic delays at stop-controlled T–intersection.

Keywords Traffic delays · Intersections · Dimensionless method · Minor road

1 Introduction

Traffic delay is one of the measures used to evaluate the performance of stop–controlled intersections. Traffic delay is commonly determined as the extra time taken in a transportation facility in comparing to that of a reference value. In this matter, it is the differentiation between the time it would take into consideration to traverse a road segment under suitable conditions and the real travel time.

M. A. Sahraei (✉) · E. Kuşkapan · M. Y. Çodur
Faculty of Engineering and Architecture, Civil Engineering Department, Erzurum Technical University, Erzurum, Turkey
e-mail: ali.sahraei@erzurum.edu.tr

M. Petrović and L. Novačko (eds.), *Transformation of Transportation*, EcoProduction,
https://doi.org/10.1007/978-3-030-66464-0_3

In general, there are two main forms of priority T-intersection, i.e. the Two-Way Stop-Controlled (TWSC) and All Ways Stop-Controlled (AWSC). The Highway Capacity Manual (HCM) (TRB 2010) described that a three-leg intersection could also be taken into consideration as a particular type of TWSC intersection, as long as the single minor road is controlled by a stop sign. Such type of intersections operates on a basis of the right-of-way of traffic movements. As such, motorists performing a right turn movement (in a case of a left-hand driving system) from a minor street have minimum right-of-way in regards to the related traffic laws in various regions. In this matter, the efficiency of a stop–controlled junction is highly affected with delay induced by low priority actions on the minor street (Wagner 1966).

This paper discusses the results of a study carried out to develop a model for the estimation of traffic delays experienced by the minor approach vehicles at stop-controlled T–intersections.

This paper is organized including Sect. 2 provides literature review related to the existing research about traffic delay prediction models. Section 3, methodology is described. Results and discussion are presented to demonstrate the benefits and usefulness in Sect. 4. Concluding statements are described in Sect. 5.

2 Literature Review

2.1 Evolution of Delay Assessment Techniques

One of the earliest studies by Weiss and Maradudin (Cohen et al. 1955) considered the delay to a single minor street vehicle arriving at a stop-controlled intersection with a single approach of major street traffic for the case of normal distributions of gap acceptance and major approach headways. The probability distribution of the delay was attained by contemplating that if a vehicle merges in a given interval, the total delay it had experienced is the total of all earlier rejected gaps. Tanner (Raff 1950), in another study, introduced a technique to evaluate the average delay to minor road vehicles at priority junctions based on an assumption that the major approach traffic has absolute priority.

In addition, Brilon (Brilon 2015) considered the issue of average delay when traffic on the major street is comprised of a combination of cars and trucks. The existence of large trucks can create moving queues of vehicles on the major street, led by one truck, under Weiss' presumptions. Under the said presumptions, the major approach flow is recognized by blocks of vehicles led by trucks, during which time no merging or crossing can occur, with blocks segregated by gaps of randomly distributed, non-queued vehicles, during which time merging may or may not occur.

Brilon (Brilon 2015) asserted that the basic presumptions discussed in Brilon (Brilon 2008) are not an appropriate method for practical utilize. Furthermore, the large number of concurrent equations pointed out in these documents may disturb the reader. Consequently, Brilon (Brilon 2015) attempted to improve previous former methods. The circumstances of applicability are more oriented on practical needs and simplified solution lastly focused.

Brilon (Brilon 2015) detailed that the description of the priority system by the M/M/1 queue method can create a bias, particularly for large main traffic volumes. Apart from this issue, it is described that the M/M/1 method are only appropriate for a process where the demand (q) and the capacity (c) are steady over time and where $q < c$ (under-saturated conditions). Both of these restrictions generally are not reasonable in real traffic situations. Consequently, solutions are needed for periods with $q > c$. In particular, Brilon (Brilon 2015) asserted that a worthwhile solution for the average delay in the time-dependent system ought to be a transition among the steady state delay (M/M/1) and the deterministic delay. This transition curve should be depending on an approximation.

Next, Caliendo (Caliendo 2014), Ma (Ma et al. 2013), and Anastasopoulos (Anastasopoulos et al. 2012) utilised the individual vehicle delay method, while the method by Weiss and Maradudin (Cohen et al. 1955), to analyse the results of non-stop gap acceptance functions. By applying the delay techniques of Weiss and Maradudin (Cohen et al. 1955), they selected a value of the critical gap so that it would produce the same predicted individual vehicle delay when compared to the non-stop gap acceptance function. They contemplated over two conditions, namely an exponential distribution and a gamma distribution of gap acceptance.

Over the past four decades, there has been a substantial amount of research done in establishing analytical techniques to analyse the capacity, delay, and queue of minor approach vehicles waiting to cross a main approach of traffic (Cheng and Allam 1992; Cvitanić et al. 2012; Rossi et al. 2013; Sahraei and Puan 2018; Sahraei and Akbari 2020b; Sahraei et al. 2018).

For the evaluation of intersections flow/delay, there are several functional computer-based techniques including Taylor (Weinert 2000), Van (Adams 1936), Idris (Idris et al. 2017), Dharmawan (Dharmawan et al. 2017), and Mahalleh (Mahalleh et al. 2017). It is obvious that the greater part of techniques, which is based on difficult applications of queuing theory, are too complicated for the reason at hand, even with simplifying presumptions of negative exponential headways and existence of steady state situations by Mohan (Mohan and Chandra 2016), Troutbeck (Troutbeck 2014), Guo (Guo et al. 2014) Tupper (Tupper et al. 2013), Wu (2012), (McGowen and Stanley 2012), (Lord-Attivor and Jha 2012), Sahraei (Sahraei and Akbari 2019), Devarasetty (Devarasetty et al. 2012).

Brilon (Brilon 1995) described that the guidelines for the analysis of delays at stop-controlled intersections apply different techniques. For example, Highway Capacity Manual (HCM)(TRB 2000, 2010) utilises the equation by Akcelik and Troutbeck. This technique is based on an estimated solution, which can be traced back to works by Kimber and Hollis (Drew 1968). Specifically, National Research Council (TRB 2000, 2010) described that a three-leg junction, such as a T-junction, can also be considered as a particular type of Two Way Stop-Controlled (TWSC) junction, giving that the minor road is operated by a stop sign. The procedure for calculation of delay in said manual is shown as (Eq. 1) as follows:

$$D = \frac{3600}{Cm,x} + 900\,T\left[\frac{Vx}{Cm,x} - 1 + \sqrt{\left[\frac{Vx}{Cm,x} - 1\right]^2 + \frac{\left[\frac{3600}{Cm,x}\right]\left[\frac{Vx}{Cm,x}\right]}{450\,T}}\right] + 5, \tag{1}$$

where;

D Control delay (sec/veh),
V_x flow rate for movement x (veh/h),
Cm,x capacity of movement x (veh/h),
T analysis time period (h)(T = 0.25 for a 15-miin period)

The capacity of a movement is given by (Eq. 3).

Studies by Akcelik et al. (Cohen et al. 1955; Greenshields et al. 1946; Polus et al. 1996) investigated three techniques for the evaluation of delay at stop-controlled intersections, i.e. (a) Highway Capacity Manual (HCM), (b) Akcelik-Troutbeck techniques based on the delay model, which was originally suggested by Troutbeck (Ashworth 1969; Herman and Weiss 1961), and (c) the delay technique utilised in SIDRA 5 package, which is based on overflow queue, queuing theory, and gap-acceptance method (Greenshields et al. 1946; Polus et al. 1996).

A research conducted by (Sahraei et al. 2012; Sahraei et al. 2014) has described two techniques, i.e. Tanner's and HCM's method, for the evaluation of delay which are associated to the size of the gap on the main street and queue length of the minor approach. It was pointed out that delays during the daytime were much longer in contrast to those during the twilight period at priority junction in suburban locations. Based on the data obtained from two intersections using a video camera, delays to the minor street increased as the volume of the main street also increased. Given this investigation, both HCM and Tanner's methods were concluded to be unsuitable for the analysis of delays at priority junctions in Malaysia.

2.2 *Dimensionless Technique*

The dimensionless analysis provides a technique that reduces complicated physical issues to the easiest form, prior to acquiring a quantitative solution. Sonin (Sonin 2001) describes that the principal application of the dimensional method is to deduce from a research of dimensions of the parameters in any physical system, particularly restrictions on the form of any possible relationship between those parameters. The technique is of excellent generality and mathematical simplicity.

Quantities values depend on approved weighing scales, which is on a system of measurement units and are also known as dimensional quantities. Quantities values are independent of the utilized system of units. Examples of dimensional quantities include time, length, energy, force, and so on (Ashworth and Bottom 1977).

In both engineering and science, dimensional evaluation is the research of the relationships among various physical quantities by determining their essential dimensions (such as length, mass, time, and so on) and units of calculate (such as kilometre per hour and sec per vehicle) in addition to tracking these dimensions as computations or comparisons that are carried out. Switching from one dimensional unit to another is generally somewhat complicated. Dimensional analysis or more particularly, the factor-label technique is also recognized as the unit-factor technique. It is a widely utilized method for conversions utilizing the regulations of algebra (Babaei et al. 2017a; Brilon 2015; Kiss et al. 2017; Kyte et al. 1996; Specht 1991; Tian et al. 2017).

A dimensionless technique is a method in which all the parameters and especially those of length, time, and mass are indicated in dimensionless form. The existing analysis is expected to clear the ground for mathematical treatments of specific issues (Abou-Henaidy et al. 1994; Avelar et al. 2017; Babaei et al. 2017b; Castillo-Santos et al. 2017). The idea of physical dimension was first introduced by Bhaskar and Nigam (Tian et al. 1999). Based on the mentioned concept, physical quantities that are commensurable have similar dimension. Conversely, if they have various dimensions, they are incommensurable.

Any physically significant formula possesses equal dimensions on the right and left sides, a property identified as "dimensional homogeneity". Checking for and ensuring this is a prevalent application of dimensional evaluation. The evaluation is also routinely utilized as a check on the plausibility of derived formulas and calculations. It is commonly applied to identify forms of physical quantities and units based on their relationship to or dependence on other units.

Subdivision of quantities towards dimensional and dimensionless is conventional. For instance, angles are specified as a dimensionless quantity, but it is recognized that angles can be calculated in degrees, radians, and fractions of a right angle, which are in various units. Therefore, the value identifying the angle would depend on the selection of the measurement unit. Thus, an angle can also be assumed as a dimensional quantity (Ashworth and Bottom 1977).

3 Methodology

3.1 Case Study

Two stop–controlled intersections in Johor, Malaysia were considered for data collection purposes, i.e. intersection at Jalan Kebudayaan/Jalan Kebudayaan 3 (Fig. 1) and the intersection at Jalan Skudai-Gelang Patah/Jalan Ronggeng 18 (Fig. 2). Both intersections are characterised by two lanes per direction on all approaches. Figure 3 indicates the lane configurations of each junction. In this study, data was collected from 9 am to 6 pm. These recording durations were taken into consideration suitable for analyzing the needed traffic variables under a variety of traffic flows. In addition,

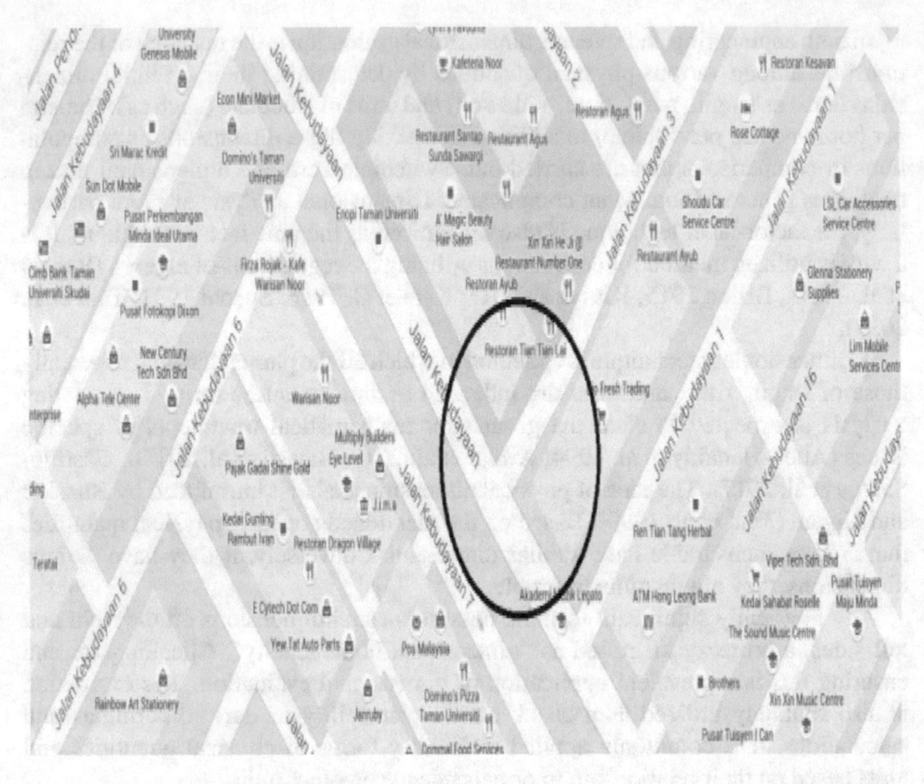

Fig. 1 Kebudayan/Jalan Kebudayaan

the purpose of selecting these two intersections was due to the preliminary short traffic counts that have confirmed reasonable volumes of turning movements which was deemed appropriate for the objectives of the current research.

This study was conducted to develop an empirical technique to calculate delays in various stop-controlled intersections. The required field data were collected using video cameras. A previous study by Ashworth (Ashworth 1976), Ke et al. (Ke et al. 2017) and Huang (Huang 2018) explained the benefits of utilising video recording technique for traffic data collection. The said technique was also utilised in several delay and gap acceptance research (Sahraei and Akbari 2020a; Teply et al. 1997). Additionally to that, the real data collected using video cameras were suitable and enough for the model development. Consequently, the simulation tools for collecting data were not utilized in this study.

In the case of data collection, each one of the recordings comprising the recorded scenes was played back numerous times to obtain the appropriate information as detailed below.

- Vehicle departure and arrival times on the minor road,
- Vehicle arrival times for main street, and
- Traffic composition.

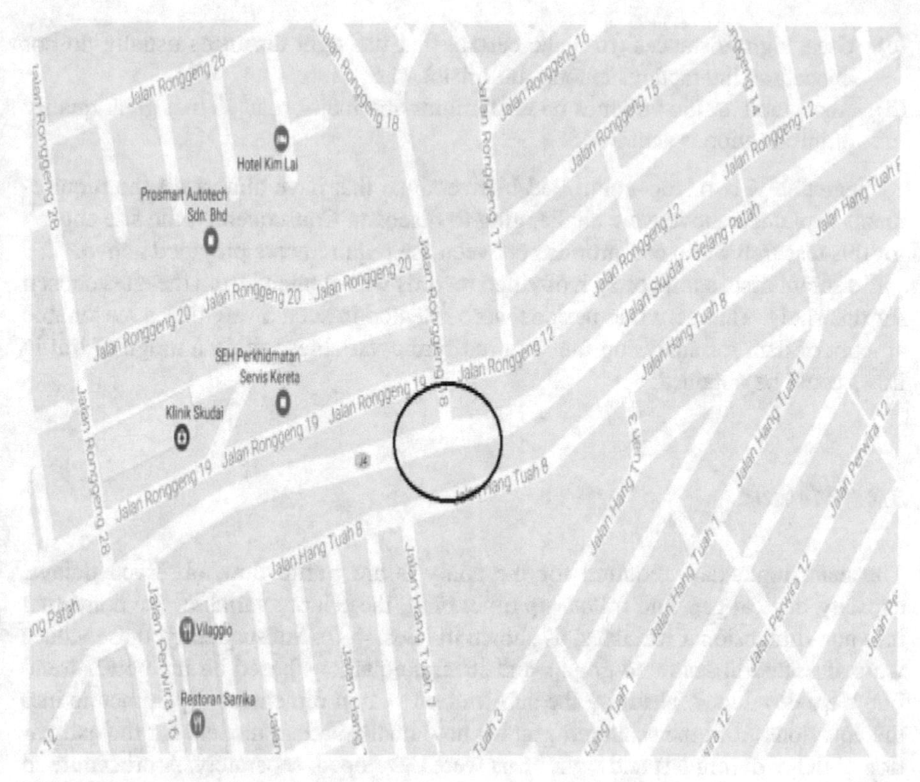

Fig. 2 Skudai-Gelang Patah/Jalan Ronggeng

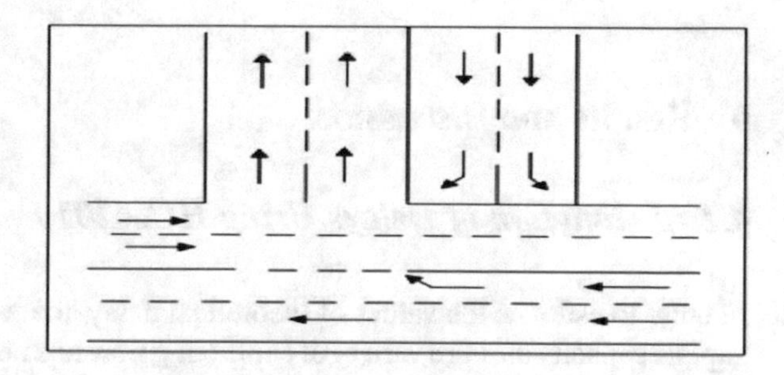

Fig. 3 Traffic lanes configuration on each junction studied

A laptop utilized to acquire the data identifying the above data from the recordings. Several visits were prepared to the different stop–controlled intersections throughout suburban and urban locations. Then, these two intentions were to recognize appropriate study area for data collection reasons. Selection on sites to be analyzed was depending on the following requirements:

(a) Great overhead advantage points for video recording reasons,
(b) Excellent safety and accessibility for the equipment throughout the data collection procedure,

(c) Great sight distances (to make certain that the sight distances usually do not impact the interactions among motorists), and
(d) Acceptable traffic volumes on each minor and major road so that great quality of information is acquired.

Regrettably, two stop–controlled intersections that have almost all the requirements explained above were challenging to discover. Consequently, the site chosen for this research was a compromise between the requirements provided above.

To summarise, a total of 18 h of video records was obtained from the sites chosen for this study. The recording periods were selected in such a way that a reasonable amount of data for analysing the required traffic variables within a range of traffic flows could be obtained.

3.2 Method

The main input data required for the analysis are traffic flow, observed delays, capacity, critical gap, and follow-up time. First, these input variables are converted into non-dimensional variables, as shown in (Eqs. 4–6). Subsequently, these values were allocated in scattered graphs and several equations based on the power-trend line were developed. Next, by the substitution of non-dimensional parameters into the equations from the scattered graphs, new mathematical models for the estimation of delay during left and right turns were developed, separately. A procedure in regards to the conversion of dimensionless values and development of new model is explained in details in the part B of the following section.

4 Results and Discussion

4.1 Estimation of Delays Using HCM2010

In order to estimate the values of theoretical delay, the values of minor road critical gap and capacity during a variety of conflicting flow rate need to be computed utilizing the formulas proposed by HCM 2010 (TRB 2010), as shown in (Eqs. 2–3). In this case, the value of base critical gap for (Eq. 2) can be found from Table 1.

$$t_{c,x} = t_{c,base} + t_{c,HV}P_{HV} + t_{c,G}G - t_{c,T} - t_{3,LT}, \quad (2)$$

where; $t_{c,x}$ = critical gap for movement x (s), $t_{c,base}$ = base critical gap from Table 1, $t_{c,HV}$ = adjustment factor for heavy vehicles (1.0 for two-lane major streets and 2.0 for four-lane major streets) (s), P_{HV} = proportion of heavy vehicles for minor movement, $t_{c,G}$ = adjustment factor for grade (s), G = percent grade divided by

Table 1 Base critical gaps and follow-up times for TWSC intersections

Vehicle movement	Base critical gap, $t_{c,base}$ (s)	
	Two-lane major street	Four-lane major street
Left turn from major	4.1	4.1
Right turn from minor	6.2	6.9
Through traffic on minor	6.5	6.5
Left turn from minor	7.1	7.5

100, $t_{c,T}$ = adjustment factor for each part of a two-stage gap acceptance process (1.0 for first or second stage; 0.0 if only one stage) (s), and $t_{3,LT}$ = adjustment factor for intersection geometry (0.7 for minor-street left-turn movement at three-leg intersection; 0.0 otherwise) (s).

$$C_{p,x} = V_{c,x} \frac{e^{-v_{c,x} t_{c,x}/3600}}{1 - e^{-v_{c,x} t_{f,x}/3600}}, \tag{3}$$

where; $c_{p,x}$ = potential capacity of minor movement x (veh/h), $v_{c,x}$ = conflicting flow rate for movement x (veh/h), $t_{c,x}$ = critical gap for minor movement x (s), and $t_{f,x}$ = follow-up time for minor movement x (s).

The values of critical gap (i.e. Equation 2) at multilane junction were computed about 3.30 and 4.20 s for left turn and right turn from minor road, respectively. For the follow-up time, the value of 2.10 s for left turning vehicles and 2.40 s for right turning vehicles were calculated which are based on the average observed data. Eventually, traffic delay (i.e. theoretical delays) was estimated using (Eq. 1), and results compared with the observed values. Figure 4 presents the variations of the observed and theoretical delays.

It can be clearly seen from Fig. 4 that there are obvious differences between the observed delays and the values estimated using the HCM2010 for both left and right turning movements. A statistical student t-test analysis was estimated to verify the significant differences among the two sets of delays as summarised in Table 2. The outcome of the t-test shows that the differences between the two sets of delays are significant. It implies that the existing delay model is not directly applicable for the computation of delay at stop-controlled intersections in Malaysia.

4.2 Model Development Based on the Dimensionless Method

To apply the dimensionless method, both input and output variables were first recognised. The input variables are the traffic flow (veh/h), follow-up time (sec), capacity (veh/h), and critical gap (sec). The output variable is the traffic delay (sec/veh).

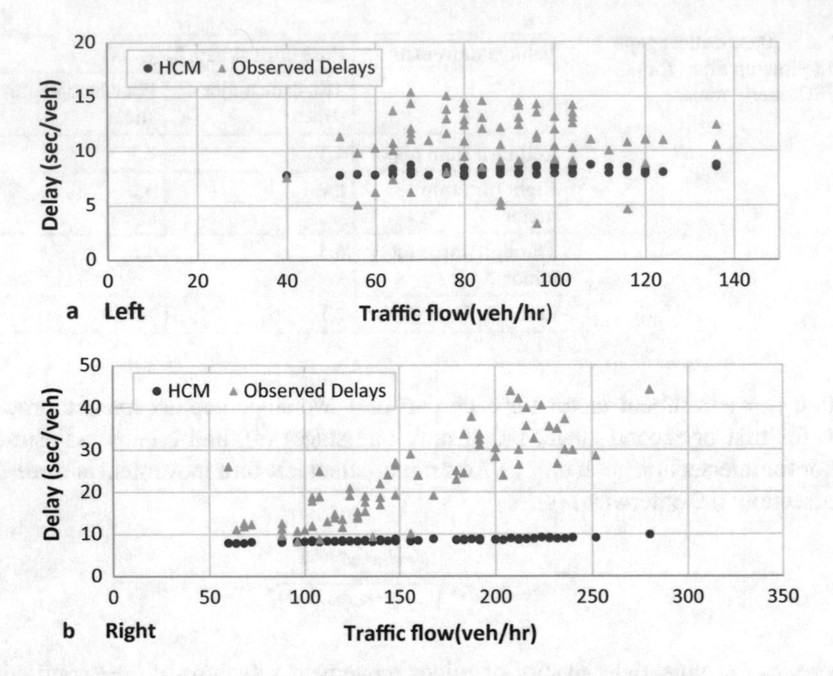

Fig. 4 Variations of the observed and HCM2010 delays, **a** Left turn, **b** Right turn

Table 2 Statistical test for HCM and observed values

Traffic movement	No. of observation	p-value
Left turn	72	$3.34e^{-10}$
Right turn	72	$1.09e^{-10}$

In the first step, the conversion of the units into non-dimensional variables was performed as follows:

$$A = \frac{\text{traffic flow}}{\text{capacity}} = \frac{\alpha}{\beta} \rightarrow \frac{\text{veh/h}}{\text{veh/h}} = 1 \tag{4}$$

$$\mathrm{B} = \frac{\text{critical gap}}{\text{follow} - \text{up time}} = \frac{\gamma}{\delta} \rightarrow \frac{\text{sec}}{\text{sec}} = 1. \tag{5}$$

In order to the making non-dimension for output variable (traffic delay), another unit was used as follows:

$$\mathrm{C} = \text{Traffic delay} * \text{traffic flow} = \psi * \alpha \rightarrow (\text{sec/veh}) * [(\text{veh/h})/3600] = 1. \tag{6}$$

Based on the above conversion, the dimensions of the variables were equalled. In the next step, variables *A, B,* and *C* were plotted on a scatter graph. Figure 5 shows

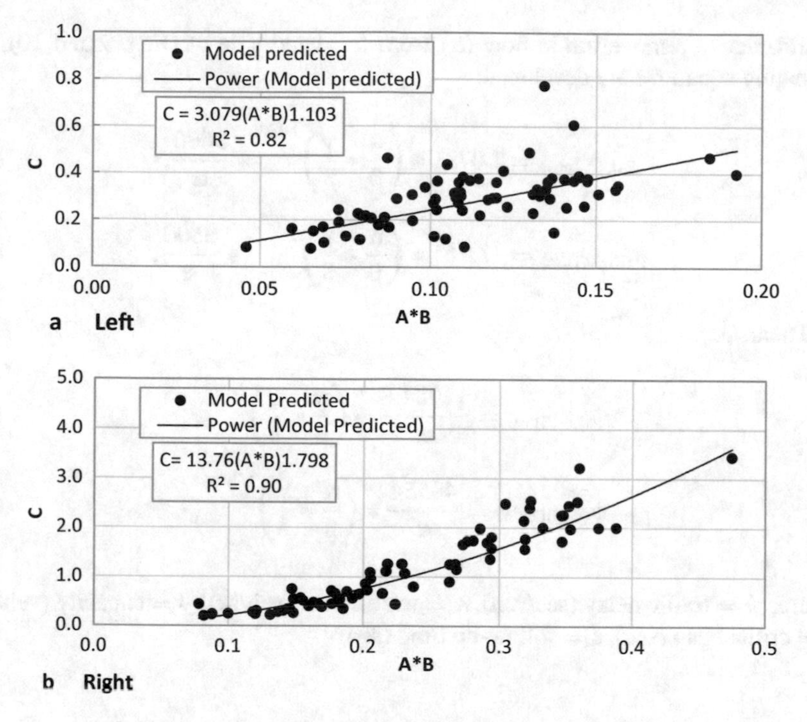

Fig. 5 Variation between inputs and output variables, **a** Left turn, **b** Right turn

the relationship between the product of delay and traffic flow, i.e. variable *C*, and the product of variables *A* and *B* for left turn and right turn movements, respectively.

Equations 7–8 can be deduced from Fig. 5a and b which are based on the power trend/regression type. Both models are regarded as appropriate because the R^2-values are relatively high, i.e. 0.82 for left turn and 0.90 for right turn movements.

$$\text{Left Turn} \rightarrow C = 3.0791 * (A * B)^{1.103}, \tag{7}$$

$$\text{Right Turn} \rightarrow C = 13.765 * (A * B)^{1.7984}, \tag{8}$$

where all variables are defined earlier. To determine the applicability of this technique, (Eqs. 4–6) is substituted into (Eqs. 7 and 8) separately, as follows:

$$\psi_{(\text{Left Turn})} * \alpha * \frac{1}{3600} = 3.0791 * \left(\frac{\alpha}{\beta} * \frac{\gamma}{\delta}\right)^{1.103}, \tag{9}$$

$$\psi_{(\text{Right Turn})} * \alpha * \frac{1}{3600} = 13.765 * \left(\frac{\alpha}{\beta} * \frac{\gamma}{\delta}\right)^{1.7984}. \tag{10}$$

In order to remove traffic flow (α) from left-hand side of (Eqs. 9 and 10), the following equations are developed:

$$\psi_{(\text{Left Turn})} = 3.0791 * \left(\frac{\alpha}{\beta} * \frac{\gamma}{\delta}\right)^{1.103} * \frac{3600}{\alpha}, \quad (11)$$

$$\psi_{(\text{Right Turn})} = 13.765 * \left(\frac{\alpha}{\beta} * \frac{\gamma}{\delta}\right)^{1.7984} * \frac{3600}{\alpha}, \quad (12)$$

Then;

$$\psi_{(\text{Left Turn})} = \frac{11084.8}{\alpha} * \left(\frac{\alpha}{\beta} * \frac{\gamma}{\delta}\right)^{1.103}, \quad (13)$$

$$\psi_{(\text{Right Turn})} = \frac{49554}{\alpha} * \left(\frac{\alpha}{\beta} * \frac{\gamma}{\delta}\right)^{1.7984}, \quad (14)$$

where: ψ = traffic delay (sec/veh), α = traffic flow (veh/hr), β = capacity (veh/hr), γ = critical gap (sec), δ = follow-up time (sec).

4.3 Model Validation

The process of validation was conducted based on the comparisons between observed data and outputs of the developed formulas (Eqs. 13 and 14) using new data from new site study (i.e. Jalan Tembaga/Jalan Tembaga Kuning1) for left- and right-turning movements from minor road. In this regard, a total number of 36 new data sets were collected. The results of the validation study on the movements can be seen in Fig. 6a and b. In general, Fig. 6 describes the variations of the observed delays and the delays predicted using (Eqs. 13 and 14) for the given input data. In this research, the actual data collected from video cameras were suitable and enough for the model validation. Consequently, the simulation tools for collecting data were not utilized. The delays to left and right turning vehicles are based on a range of conflict flows of 128–306 veh/h and 266–588 veh/h, respectively. In general, the models show that the delay to minor road vehicles increases as the traffic demand increases as expected.

A visual representation (Fig. 7) of the relative quality of the predicted delays for both left and right turning vehicles was also assessed using a 45-degree line plot. As can be seen from Fig. 7, the observed delays to the minor road vehicles scattered around the predicted delays using the models given in (Eqs. 13 and 14). Visually, Fig. 7 indicates that the predicted delays are in a good agreement with the observed values.

Table 3 provides the values of Residual Sum of Squares (RSS), Mean-Square Error (MSE) and Root-Mean-Square Error (RMSE), for all movements. The RMSE

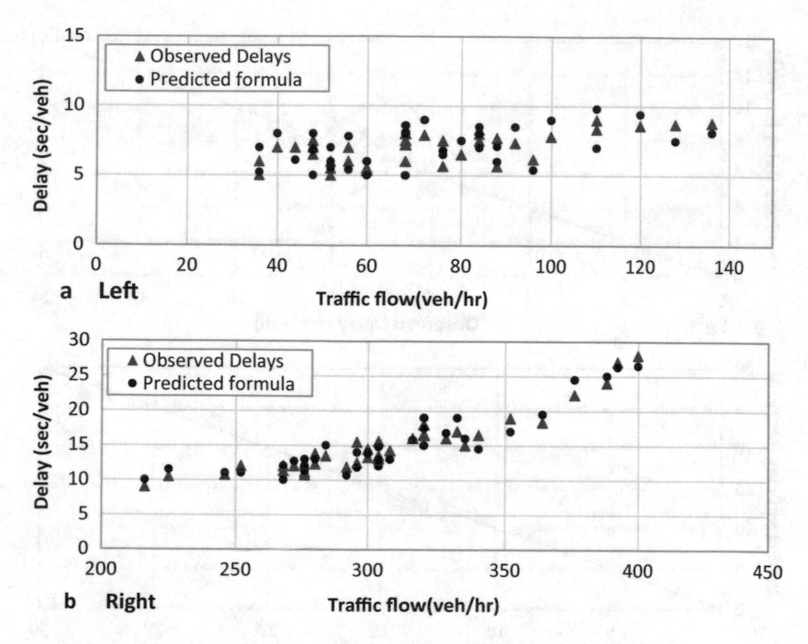

Fig. 6 Variations of the predicted and observed delays, **a** Left turn, **b** Right turn

and MSE is a regularly computes of the differences between actual data set and the values computed by the formulation. Particularly the values closer to zero are significantly better, where the values of RMSE for right- and left-turning manoeuvre from minor road were estimated around 0.818, and 0.581, respectively. In addition, MSE is computed around 0.669 for right-turning movement and 0.337 for left-turning movements from minor road.

The RSS also known as the Sum of Squared Errors of prediction (SSE). It is to examine of the discrepancy between the actual data set and outcomes of predicted models. Similarly, a small RSS suggests a well fit of the models to the observed data. In this case, the values of RSS for right- and left-turning manoeuvre from minor road were estimated around 1.318, and 1.031, respectively.

To confirm the comparisons between the predicted and observed values, student t-Test was conducted and result is summarised in Table 3. The p-values show that the differences between the predicted and observed values are not significant. In brief, the comparisons indicate that the delays predicted using the dimensionless-based models are in a good agreement with the observed delays for a range of traffic flows.

4.4 Sensitivity Analysis

A sensitivity analysis is a method utilized to identify exactly how distinct values of an independent variable influence a certain dependent variable under a provided set

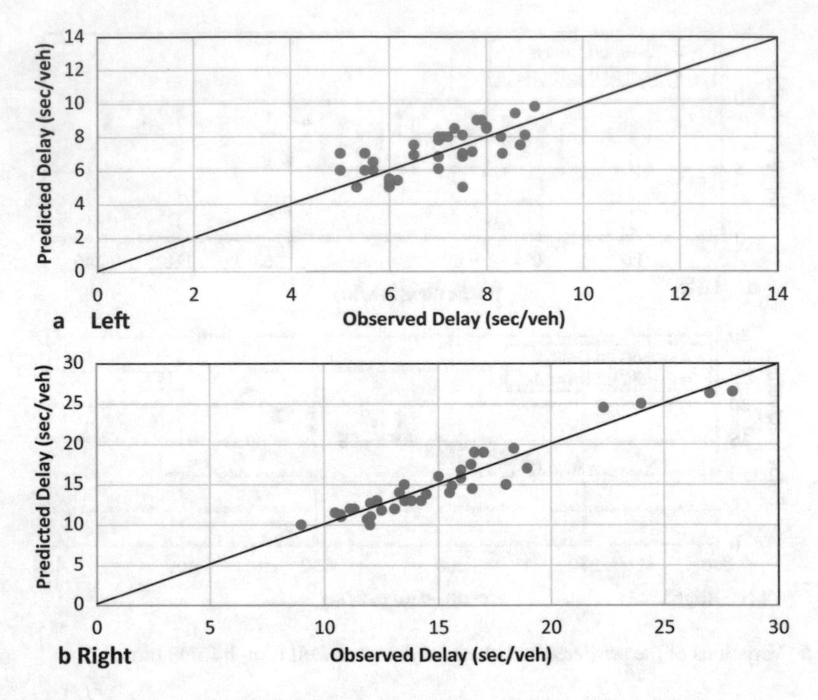

Fig. 7 Comparisons between the predicted and observed delays, a Left turn, b Right turn

Table 3 Statistical test for developed model

Traffic movement	No. of observation	p-value	RMSE	MSE	RSS
Left turn	36	0.641	0.581	0.337	1.031
Right turn	36	0.989	0.818	0.669	1.318

of presumptions. This method is utilized within particular boundaries that rely on one or more input variables.

Based on the models developed in this study, it apparent that independent variables (i.e. input parameters) had several effects on the results of the expected traffic delays. In this regard, by increasing the proportions of right-turning flow rate on the minor road from 10% to 50%, the values of delays would increase amount 9 (sec/veh), as shown in Fig. 8a. In terms of effects of capacity, the values of traffic delay are expected to decrease almost 11 (sec/veh) when capacity values of right-turning on the minor road are increased by 10–50%. Accordingly, by increasing the proportions of right-turning flow rate on the minor road to the 10, 20, 30, 40, and 50%, the values of delays would increase 23.90, 25.62, 27.31, 28.97,30.61 (sec/veh), as shown in Fig. 8a. In terms of effects of capacity, the values of traffic delay are expected to decrease to the 18.66, 15.96, 13.82, 12.09, and 10.68 (sec/veh) when the capacity of right-turning lane on the minor road is increased in the same percentages.

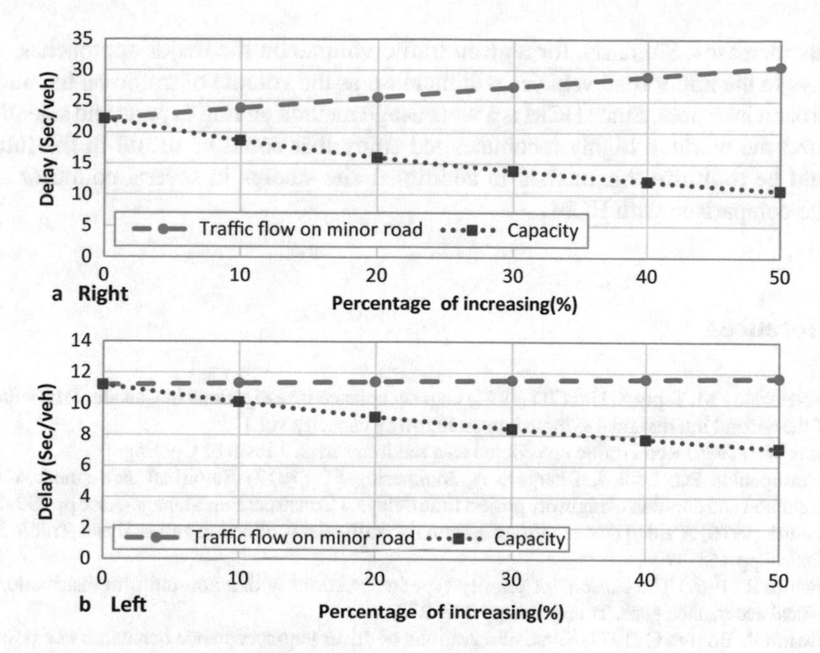

Fig. 8 Results of sensitive analysis, **a** Right turn, **b** Left turn

On the other hand, as shown in Fig. 8b, if the left-turning flow rate on the minor road is increased by 10–50%, the traffic delay is expected to increased amount 2 (sec/veh). If the capacity of left-turning lane on the minor road increased by 10–50%, traffic delay is expected to decrease around 4 (sec/veh). Accordingly, if the left-turning flow rate on the minor road is increased by 10, 20, 30, 40, and 50%, the traffic delay is expected to increased 11.32, 11.42, 11.53, 11.61, and 11.70 (sec/veh). However, if the capacity of left-turning lane on the minor road increased in the same percentages, traffic delay is expected to decrease 10.09, 9.17, 8.39, 7.73, and 7.17 (sec/veh).

5 Conclusion

This paper described the application of a dimensionless method for the analysis of traffic delays at stop-controlled intersections. Two mathematical models for left- and right-turning vehicles on their respective turning lanes on the minor road were developed. The outcomes of this investigation revealed that the delays predicted using the developed models were in a good agreement with the observed delays. Such a finding suggests that the dimensionless-based models are applicable for the analysis of delays at stop-controlled intersections in Malaysia.

Both models show that, for a given traffic demand on the minor approach, delays to the minor road vehicles will increase as the volume of the traffic on the main approach roads increases. Similarly, for a given traffic volume on the major approaches, the delays to the minor road vehicles will increase as the volume of traffic on the minor approach increases. Since HCM is a wide usage method among experts and scientists around the world, a highly recommended study that could be useful in the future would be to utilize this method in additional site studies in several countries and make comparison with HCM.

References

Abou-Henaidy M, Teply S, Hunt JD (1994) Gap acceptance investigations in Canada. Proceedings of the second international symposium on highway capacity vol 1

Adams WF (1936) Road traffic considered as a random series. J Instn of Civ Engrs

Anastasopoulos PC, Labi S, Bhargava A, Mannering FL (2012) Empirical assessment of the likelihood and duration of highway project time delays. J Construct Eng Manage-Asce pp 390–398

Ashworth (1976) A video tape recording system for traffic data collection and analysis. Traffic Eng Control pp 468–470

Ashworth R (1969) The capacity of priority-type intersections with a non-uniform distribution of critical acceptance gaps. Transp Res pp 273–278

Ashworth R, Bottom C (1977) Some observations of driver gap-acceptance behaviour at a priority intersection. Traffic Eng Control pp 569–571

Avelar AHdF, Canestri JA, Bim C, Silva MG, Huebner R, Pinotti M (2017) Quantification and analysis of leaflet flutter on biological prosthetic cardiac valves. Artif Organs pp 835–844

Babaei H, Mirzababaie Mostofi T, Armoudli E (2017a) On dimensionless numbers for the dynamic plastic response of quadrangular mild steel plates subjected to localized and uniform impulsive loading. Proceedings of the Institution of Mechanical Engineers, Part E: J Proc Mech Eng pp 939–950

Babaei H, Mostofi TM, Alitavoli M, Saeidinejad A (2017b) Experimental investigation and dimensionless analysis of forming of rectangular plates subjected to hydrodynamic loading. J Appl Mech Tech Phys pp 139–147

Brilon W (1995) Methods for measuring critical gap. Ruhr-University, Bochum, Germany

Brilon W (2008) Delay at unsignalized intersections. Transp Res Rec pp 98–108

Brilon W (2015) Average delay at unsignalized intersections for periods with variable traffic demand. Transp Res Rec: J Transp Res Board pp 57–65

Caliendo C (2014) Delay time model at unsignalized intersections. J Transp Eng p 04014042

Castillo-Santos K, Ruiz-López I, Rodríguez-Jimenes G, Carrillo-Ahumada J, García-Alvarado M (2017) Analysis of mass transfer equations during solid-liquid extraction and its application for vanilla extraction kinetics modeling. J Food Eng pp 36–44

Cheng TCE, Allam S (1992) A review of stochastic modelling of delay and capacity at unsignalized priority intersections. Eur J Oper Res pp 247–259

Cohen J, Dearnaley E, Hansel C (1955) The risk taken in crossing a road. OR. pp 120–128

Cvitanić D, Breški D, Vidak B (2012) Review, testing and validation of capacity and delay models at unsignalized intersections. PROMET-Traffic and Transportation pp 71–82

Devarasetty P, Zhang Y, Fitzpatrick K (2012) Differentiating between left-turn gap and lag acceptance at unsignalized intersections as a function of the site characteristics. J Transp Eng pp 580–588

Dharmawan A, Ashari A, Putra AE (2017) Mathematical modelling of translation and rotation movement in quad tiltrotor. Int J Adv Sci Eng Inf Technol pp 1104–1113

Drew DR (1968) Traffic flow theory and control

Greenshields BD, Schapiro D, Ericksen EL (1946) Traffic performance at urban street intersections

Guo R-J, Wang X-J, Wang W-X (2014) Estimation of critical gap based on Raff's definition. Comput Intell Neurosci p 16
Herman R, Weiss G (1961) Comments on the highway-crossing problem. Oper Res pp 828–840
Huang T (2018) Traffic speed estimation from surveillance video data. Proceedings of the IEEE Conference on Computer Vision and Pattern Recognition Workshops, Salt Lake City, USA
Idris N, Hashim SZM, Samsudin R, Ahmad NBH (2017) Intelligent learning model based on significant weight of domain knowledge concept for adaptive e-learning. Int J Adv Sci Eng Inf Technol pp 1486–1491
Ke R, Pan Z, Pu Z, Wang Y (2017) Roadway surveillance video camera calibration using standard shipping container. 2017 International smart cities conference (ISC2), IEEE, Wuxi, China
Kiss D, Duretz T, Podladchikov Y, Schmalholz SM (2017) Dimensional analysis and scaling for shear zones caused by thermal softening and comparison with 2D numerical models. EGU General Assembly Conference Abstracts
Kyte M, Tian Z, Mir Z, Hameedmansoor Z, Kittelson W, Vandehey M, Robinson B, Brilon W, Bondzio L, Wu N (1996) Capacity and level of service at unsignalized intersections. Final Report. vol 2. All-Way Stop-Controlled Intersections
Lord-Attivor R, Jha MK (2012) Modeling gap acceptance and driver behavior at stop controlled (priority) intersections in developing countries. Proceedings of the 6th WSEAS international conference on Computer Engineering and Applications, and Proceedings of the 2012 American conference on Applied Mathematics, World Scientific and Engineering Academy and Society (WSEAS)
Ma D-F, Ma X-L, Jin S, Sun F, Wang D-H (2013) Estimation of major stream delays with a limited priority merge. Can J Civ Eng pp 1227–1233
Mahalleh VBS, Selamat H, Sandhu F (2017) Review on psychological crowd model based on lebon's theory. TELKOMNIKA (Telecommunication Computing Electronics and Control)
McGowen P, Stanley L (2012) Alternative methodology for determining gap acceptance for two-way stop-controlled intersections. J Transp Eng pp 495–501
Mohan M, Chandra S (2016) Review and assessment of techniques for estimating critical gap at two-way stop-controlled intersections. Eur Transport-Trasporti Europei
Polus A, Craus J, Reshetnik I (1996) Non-stationary gap acceptance assuming drivers' learning and impatience. Traffic Eng Control
Raff MS (1950) A volume warrant for urban stop signs
Rossi R, Meneguzzer C, Gastaldi M (2013) Transfer and updating of Logit models of gap-acceptance and their operational implications. Transp Res Part C: Emerg Technol pp 142–154
Sahraei MA, Akbari E (2019) Implementing the equilibrium of probabilities to measure critical gap at priority junctions. J Test Eval
Sahraei MA, Akbari E (2020a) Effect of motorcycle on the critical gap at priority junctions. Aust J Civ Eng pp 1–13
Sahraei MA, Akbari E (2020b) Review and evaluation of methods for estimating delay at priority junctions. Aust J Civ Eng pp 1–14
Sahraei MA, Che Puan O, Yasin MA (2014) Minor road traffic delays at priority junctions on low speed roads in Suburban Areas. Jurnal Teknologi
Sahraei MA, Puan O (2018) Traffic delay estimation using artificial neural network (ANN) at unsignalized intersections. The 3rd International Conference on Civil, Structural and Transportation Engineering (ICCSTE'18), Canada
Sahraei MA, Puan OC, Hosseini SM, Almasi MH (2018) Establishing a new model for estimation of the control delay at priority junctions in Malaysia. Cogent Eng p 1424679
Sahraei MA, Puan OC, Jahandar N (2012) Comparison of delays between twilight time and day time. J Basic Appl Sci Res pp 9649–9657
Sonin AA (2001) The physical basis of dimensional analysis. Department of Mechanical Engineering, MIT, Cambridge, MA
Specht DF (1991) A general regression neural network. Neural Networks, IEEE Transactions on. pp 568–576

Teply S, Abou-Henaidy MI, Hunt JD (1997) Gap acceptance behaviour: aggregate and logit perspectives: Part 1. Traffic Eng Control pp 474–482
Tian G, Lee SL, Yang X, Hong MS, Gu Z, Li S, Fisher R, O'Connor TF (2017) A dimensionless analysis of residence time distributions for continuous powder mixing. Powder Technol pp 332–338
Tian Z, Vandehey M, Robinson BW, Kittelson W, Kyte M, Troutbeck R, Brilon W, Wu N (1999) Implementing the maximum likelihood methodology to measure a driver's critical gap. Transp Res Part A: Policy Pract pp 187–197
TRB (2000) Highway capacity manual. Transportation Research Board (TRB), National Research Council, Washington D.C
TRB (2010) Highway capacity manual. Transportation Research Board (TRB), National Research Council, Washington D.C
Troutbeck R (2014) Estimating the mean critical gap. Transp Res Rec: J Transp Res Board pp 76–84
Tupper S, Knodler Jr MA, Fitzpatrick C, Hurwitz DS (2013) Estimating critical gap—a comparison of methodologies using a robust, real-world data set. 92nd Annual Meeting of the Transportation Research Board, Washington DC
Wagner Jr FA (1966) An evaluation of fundamental driver decisions and reactions at an intersection. Highway Research Record
Weinert A (2000) Estimation of critical gaps and follow-up times at rural unsignalized intersections in Germany. Fourth international symposium on highway capacity
Wu N (2012) Estimating distribution function of critical gaps at unsignalized intersections based on equilibrium of probabilities. Transp Res Rec

In-Depth Evaluation of Reinforcement Learning Based Adaptive Traffic Signal Control Using TSCLAB

Daniel Pavleski, Mladen Miletić, Daniela Koltovska Nečoska, and Edouard Ivanjko

Abstract Adaptive Traffic Signal Control (ATSC) is today widely applied for managing traffic on signalized intersections due to its capability to reduce congestion. ATSC changes the signal programs in real-time according to the measured current incoming traffic flows when ATSC is applied. This results in an improvement in the throughput of urban networks. However, prior to the implementation of such systems they be evaluated. Evaluation of the effectiveness of complex ATSC is still a challenge and presents an open problem. For the evaluation, different measures of effectiveness to gather in-depth insight into the traffic situations of the controlled signalized intersection are needed. In this paper, an augmented version of the previously developed MATLAB based tool TSCLab (Traffic Signal Control Laboratory) is applied to evaluate a newly proposed ATSC based on self-organizing maps and reinforcement learning. The performance of the mentioned ATSC is evaluated using appropriately chosen measures of effectiveness obtained in real-time using a microscopic simulation environment based on VISSIM and a realistic traffic scenario. Obtained simulation results reveal that ATSC uses shorter phase and cycle duration, achieving a lower green time utilization but also shorter queue lengths, thus improving the throughput of the analyzed intersection compared to the existing fixed-time signal control.

D. Pavleski
Faculty of Technical Sciences, Division of Traffic and Transport, Mother Theresa University Skopje, Skopje, Republic of North Macedonia
e-mail: daniel.pavleski@unt.edu.mk

M. Miletić (✉) · E. Ivanjko
Faculty of Transport and Traffic Sciences, Department of Intelligent Transportation Systems, University of Zagreb, Zagreb, Croatia
e-mail: mladen.miletic@fpz.unizg.hr

E. Ivanjko
e-mail: edouard.ivanjko@fpz.unizg.hr

D. K. Nečoska
Faculty of Technical Sciences, Department of Traffic and Transport, St. Kliment Ohridski University, Bitola, Republic of North Macedonia
e-mail: daniela.koltovska@tfb.uklo.edu.mk

M. Petrović and L. Novačko (eds.), *Transformation of Transportation*, EcoProduction,
https://doi.org/10.1007/978-3-030-66464-0_4

1 Introduction

For the optimal use of the existing road infrastructure in cities, Adaptive Traffic Signal Control (ATSC) is applied to signalized intersections prone to significant changes in traffic demand. ATSC has an advantage over traditional fixed time signal programs as it can adapt to changes in traffic demand by changing the cycle/phase duration or by changing the phase sequence. Such changes are done according to the current traffic demand values. Today learning-based ATSC systems from the domain of Intelligent Transport Systems (ITS) are increasingly being used (El-Tantawy et al. 2013). Such systems can learn different control policies capable of solving daily, seasonal and incident caused variations in traffic flow improving the Level of Service (LoS) for all traffic entities in the road network or just specific ones (Public Transportation (PT), emergency vehicles, transit traffic) in the long term (Michailidis et al. 2018).

Evaluation of ATSC systems must be performed using realistic scenarios prior to their implementation in an urban environment. This is important to assess the possible improvement of LoS and to perform setup of control parameters for adaptation. Also, potential problems in the adaptation of the signal programs must be identified and corrected. The area of ATSC evaluation is still open, as many different ATSC systems are available, including different Measures of Effectiveness (MoEs) for evaluation (Wahlstedt 2013). One of the approaches for this is the application of the framework TSCLab (Traffic Signal Control Laboratory) developed by the authors where the microscopic traffic simulator VISSIM in combination with the evaluated ATCS is applied (Pavleski et al. 2020). TSCLab can gather different MoEs and evaluate them prior to real-world implementation, ensuring a good insight into the functioning of the ATCS. In this paper, the existing version of TSCLab is augmented to cope with more complex signal programs and used to evaluate a newly Self-Organizing Maps (SOM), and Reinforcement Learning (RL) based ATSC also developed by the authors (Miletić et al. 2020). This newly developed ATSC is denoted with SOM-RL in continuation of the paper. Established evaluation framework gives an in-depth insight into ATSC behavior regarding the chosen traffic scenario revealing potential problems and explaining how the adaptation of the signal program improves analyzed MoEs on a particular intersection.

This paper is organized as follows. After the introduction section, measures on how to analyze the performance of ATSC are described in the second section. The third section elaborates on the evaluation framework TSCLab and the following fourth section explains the evaluated traffic signal control approach. The fifth section presents the obtained simulation results, including a discussion about them. The last section ends the paper with a conclusion and future work description.

2 Performance Analysis of Adaptive Traffic Signal Control

As mentioned, in order to improve traffic flow performances, ATSC systems use real-time traffic data for optimizing signal timing parameters. As one of the essential ITS services, they have been widely deployed in urban areas for traffic control around the globe. They entered in our society as a successful concept due to their positive impacts on the mitigation of traffic congestion. To quantify the impact of ATSC on traffic conditions, an expanded set of MoEs should be determined, and the defined measures should reflect real-time "responses" of adaptive control to the traffic process.

The available literature describes various MoEs proposed by the transportation practitioners included in the operational process of the traffic networks and by academia. For this purpose, research mostly focused on technological issues and the development of new tools for evaluation of an expended set of MoE using analytical and simulation techniques (Gettman et al. 2013; Day et al. 2014, 2015; Dakic et al. 2018). Most of the mentioned research has been used as a starting point towards a choice of MoEs for the development of the mentioned framework TSCLab, a MATLAB based tool for monitoring and evaluation of ATSC (Pavleski et al. 2020). Table 1 provides a summary of the in this paper selected measures used to evaluate the SOM-RL based ATSC.

Following the contents in Table 1, a definition of each measure follows:

Cycle length is the time required for one full cycle of signal indications and green time duration is the length of the green interval; Maximum green time utilization ratio is calculated as a ratio of used green time to available maximum green time;

Arrived vehicles per cycle is the number of arrived vehicle/signal cycle; Served vehicles per green signal is the number of served vehicles per green signal/signal cycle;

Green/Red occupancy ratio is a measure of capacity utilization that uses the amounts of detector occupancy during the green and red portion of a phase; Queue length is the number of vehicles in the queue; Delay is a measure for the stopped time per vehicle; Stops is the number of vehicles stops;

Percent of arrived vehicles on green signal is the percent of total vehicles arriving during a given signal cycle/green.

Table 1 Selected performance measures for evaluation of RL based ATSC (Pavleski et al. 2020)

Tasks	Measures for performance evaluation
Signal timing	Cycle length and Green time duration
	Maximum green time utilization ratio
Throughput	Arrived vehicles per cycle
	Served vehicles per green signal
Capacity	Green/Red occupancy ratio
	Queue length/per cycle
	Stops
Progression quality	Percent of arrived vehicles on green signal

All presented measures are collected and processed automatically by TSCLab. Thus, the post-simulation in-depth analysis of chosen MoEs can be done in TSCLab. This alleviates the evaluation process, unlike current approaches that require manual processing of gathered simulation data. More details about the evaluated SOM-RL based adaptive control strategy are described in continuation.

3 TSCLab Evaluation Framework

Evaluation of the impact of the signal operations of ATSC is an important and challenging research issue that requires a particular set of tools, resources, and expertise. Adequate performance analysis that provides enough information is crucial for the actual quality of responsiveness of ATSC. To gather and visualize a set of relevant data, which describe the performances of ATSC in real-time, a new tool named as TSCLab has been developed. It was successfully applied to evaluate the effectiveness of UTOPIA/SPOT ATSC at an isolated signalized urban intersection on a road network in Skopje (Pavleski et al. 2020).

As mentioned above, this paper is a continuation of the author's previous work and focuses on the performance evaluation of two types of signal control methods: Fixed Time Signal Control (FTSC) and SOM-RL based ATSC using an augmented version of TSCLab. Both traffic control methods were tested on a complex urban intersection located on the broader center area of the city of Zagreb using the VISSIM microscopic simulator. To provide an in-depth analysis of the evaluated signal control strategies' effectiveness, the emphasis on improving and extending the existing TSCLab was on adding functionality to enable evaluation of more complex signal programs containing several phases and signal groups for managing complex urban intersections.

The simulation framework for FTSC and SOM-RL evaluation and testing is displayed in Fig. 1. It consists of the TSCLab as a MATLAB based tool and microscopic traffic simulator VISSIM. As a real-time evaluation tool, TSCLab reads data from VISSIM objects, processes them and visualizes the relevant data which describe the MoEs. This means that TSCLab does not impact on the traffic signal control strategies. TSCLab has been upgraded with new functionalities, and the new augmented version (TSCab 2.0) includes the following major features:

Simultaneously accessing and reading of data from 10 different VISSIM objects of the same type. For each type of VISSIM object defined in the simulation model, a maximum of 10 different objects, e.g., a set of signal groups can be specified in TSCLab and evaluated. Figure 2 shows the main graphical user interface of the new augmented version;

During the simulation execution the obtained results are displayed in the result output window after each signal cycle;

Every simulation second is recorded and stored in the signal timing window at the end of each signal cycle. This provides an opportunity for later analysis of the results for a certain period of the day;

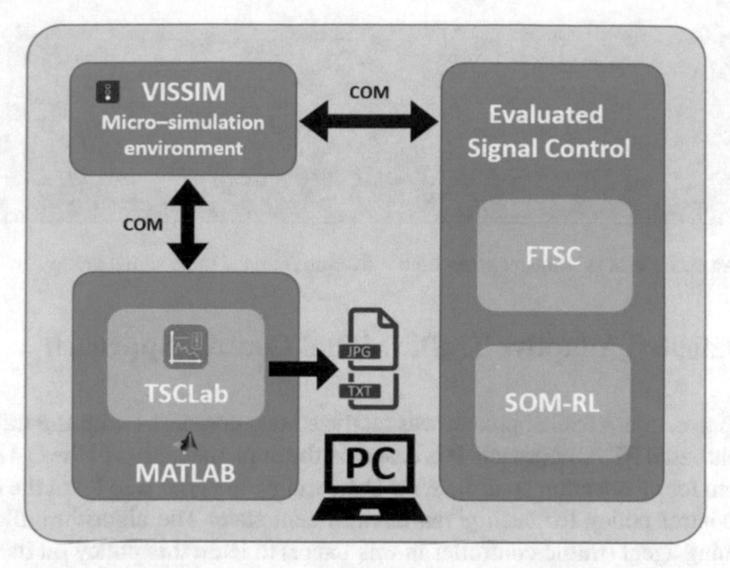

Fig. 1 Simulation framework for evaluation of SOM-RL and FTSC signal control

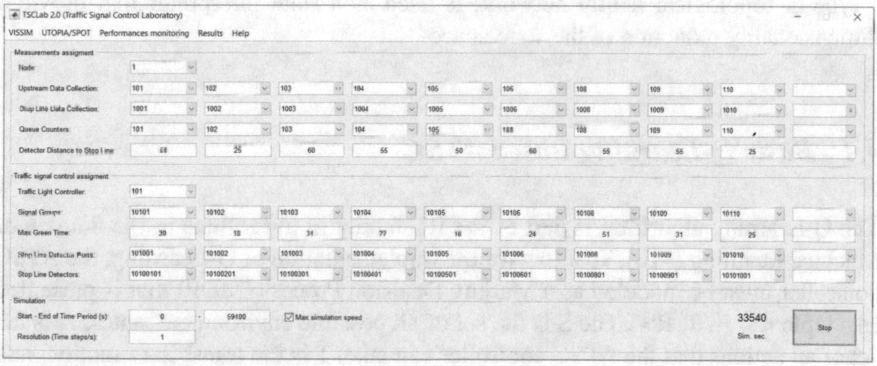

Fig. 2 Main graphical user interface of TSCLab 2.0

In order to resolve the problem of unusual subsequent activation of a green time from the same signal group the following steps were taken: (i) The mentioned activation is recorded as a signal cycle in that only one signal phase/stage was served, and (ii) The duration of green time for that particular signal group is recorded as long as it lasted; for the other signal phases/stages that were not served, TSCLab sets zero values. This implies that TSCLab 2.0 can identify so-called situations of unusual behavior of the signal groups denoting possible problems in the signal programs. Thus, enabling practitioners to correct the signal programs prior a real-world implementation. Figure 3 presents an example of subsequent activation of the green time of the same signal group that occurred during the simulation process.

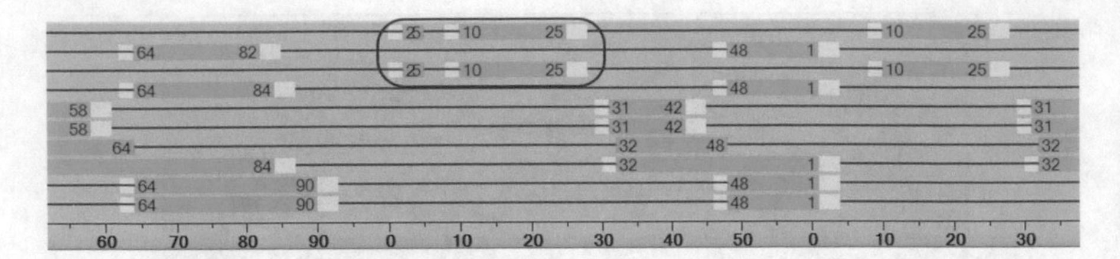

Fig. 3 An example of subsequent green time activation from the same signal group

4 Evaluated Adaptive Traffic Signal Control Approach

In this paper, one ATSC approach was analyzed and compared with the traditional timetable based FTSC approach. It is based on the implementation of the Q-Learning algorithm for intersection control. With Q-Learning, the goal is to learn the optimal signal control policy for each given environment state. The algorithm allows the Q-Learning agent (traffic controller in this paper) to learn this policy on its own by receiving a reward or penalty from the environment. This basic Q-Learning approach has been augmented with the use of SOM. In this approach, the SOM, which is a type of topological neural network, is used as a state interpreter that provides dimensionality reduction of the state space.

4.1 Basic Q-Learning Based ATSC

The Q-Learning algorithm is part of the RL family of algorithms in the domain of Machine Learning (ML). To allow for the implementation of Q-Learning, the ATSC controller must be modeled as a Markov Decision Process (MDP) that is presented as a tuple <S, A, T, R> . The S is the set of all possible environment states, A is the set of all actions that the ATSC controller can take, T is the transition function, and R is the reward function of the environment (Mannion et al. 2016). If the sets S and A have a discrete number of elements it is possible to learn an optimal control policy using the Q-Learning update function:

$$Q(s_t, a_t) \leftarrow Q(s_t, a_t) + \alpha\left(r_{t+1} + \gamma \max_{a' \in A} Q\left(s_{t+1}, a'\right) - Q(s_t, a_t)\right), \quad (1)$$

where $Q(s_t, a_t)$ is the Q value for state s and action a, α is the learning rate used to determine to what extent can the Q values be updated, r_{t+1} is the reward given from the environment one step after the action was taken, and γ is the discount factor used to provide insight into possible rewards in the next time step of the future control period.

Since the traffic environment is non-deterministic and since there is significant environment noise present in the real-time traffic state measurement, the factor α needs to be set to a smaller value to reduce the effect of noise on the learning policy. This will, in turn, require more learning iterations to achieve convergence to an optimal policy. When Q-Learning is applied to ATSC, the reward r_{t+1} is modeled to give incentive to the control when an action that improved the overall traffic situation is selected. In a more general view, the reward function should promote the desired ATSC criteria function. Another important factor to consider is state representation or the set of states S. The selection of variables that define the state space should be made to provide the best representation of the traffic state. The problem with this is that the possible number of state-action combinations exponentially increases when adding additional variables to present the state space. Therefore, the selection of appropriate state definition variables is of utmost importance to properly represent the traffic state, but not to hinder the learning process. When Q-Learning is applied in ATSC problems, the usual state definition variables consist of queue lengths on intersection approaches since that data can be easily obtained, but results in large state-action complexity (Touhbi et al. 2017).

4.2 *Q-Learning with Self-Organizing Map Dimensionality Reduction*

While Q-Learning can provide good results for ATSC, its main drawback is the exponential rise in state-action complexity with the addition of new state definition variables (Miletić et al. 2020). To cope with the increasing state-action complexity, several SOMs can be used to provide dimensionality reduction. SOM is a type of neural network that does not contain an output layer and is topological in nature, meaning that the neuron weights represent the coordinates of a neuron in the defined state space S. Neurons in the computational layer of a SOM network are usually arranged in a rectangular or a hexagonal grid with interconnections between neurons. Since SOM is mostly used in data clustering problems, the basic idea of applying it to ATSC is to arrange neuron weights in such a way that would position the neurons in most representative states in the state space S. By doing this the state space is split into n discrete segments, where n is the number of neurons in the computational layer of SOM.

SOM learning algorithm takes the input signal $X = (x_1, x_2, x_3, \cdots, x_m)$ consisting of m input variables that represent the state space S and then seeks to identify the neuron, which has the shortest Euclidean distance from the input signal X. This identified neuron sometimes referred to as the winning neuron or the Best Matching Unit (BMU), will then have its weights changed to be more similar to the input signal X. Neurons connected to the BMU will have their weights changed according to:

$$W_i(k+1) = W_i(k) + \theta\alpha_{SOM}[X(k) - W_i(k)], \tag{2}$$

where W_i is the vector of weights of a neuron i, θ is the neighborhood function, α_{SOM} is the learning rate, and k is the time step. The neighborhood function is used to determine how does the distance of a neuron from the BMU affects the change in neuron weights. The general rule is that neurons that are far away from the BMU will their weights changed only slightly, while the neurons close to the BMU will change their weights significantly. The neighborhood function is defined as:

$$\theta = \exp\left(-\frac{d^2}{2P^2}\right), \tag{3}$$

where d is the distance of a neuron to the BMU, and P is the parameter used to calibrate SOM learning. Usually, the parameter P is decreased during the learning process. By doing this initially, most neurons will react to an input signal, but as the learning continues, only a few neurons will have their weights changed. When the learning is complete, each neuron will represent a segment of the state space, which means that the initial continuous state space S is now reduced to several discrete segments.

Due to its nature to preserve the topological nature of the state space, SOM can be used as a state interpreter in the process of Q-Learning for the development of an ATSC controller, as shown in Fig. 4. The proposed SOM-Q-Learning model requires the SOM network to be trained in an off-line fashion according to the previous historical data. Once the SOM network is trained, it is used to process high dimensional

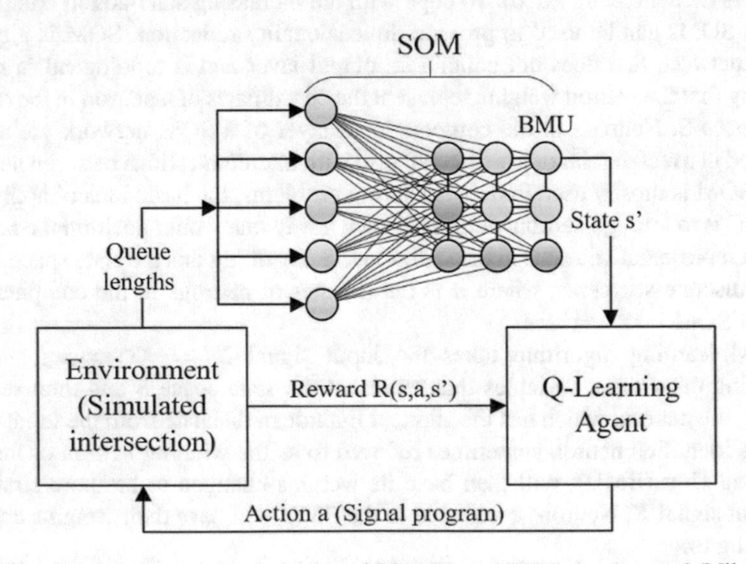

Fig. 4 Simulation framework for evaluation of SOM-RL and FTSC signal control (Miletić et al. 2020)

data representing the current state of the environment (e.g., Queue lengths on each lane approaching the intersection). When given an input signal, the identified BMU will then correspond to an identified state in the Q-Learning algorithm, which will then continue in the usual way by selecting actions and receiving rewards from the environment. The benefit of doing this is a significant decrease in the state-action complexity to improve the convergence of the applied Q-Learning algorithm (Miletić et al. 2020).

5 Use Case and Evaluation Results

To evaluate the proposed SOM-RL ATSC approach using TSCLAB, a use case is created, and the proposed traffic control approach is compared to the conventional FTSC approach.

5.1 Simulation Model and Traffic Data

The traffic simulation model used was created using the PTV VISSIM microscopic simulator. The model consists of an isolated signalized intersection modeled with real traffic data collected at the intersection of King Zvonimir street—Heinzelova street in the center of the City of Zagreb, Croatia (Vujić 2013; Miletić 2019) The intersection is characterized by significant daily changes in the transport demand, most notably during the morning (08:30 AM–09:30 AM) and afternoon (04:00 PM–05:00 PM) peak hours. The intersection is operated using the existing FTSC regime consisting of four different signal programs, as shown in Table 2. The currently used FTSC is an up-to-date regime created with the operational criteria being the reduction of control delay. The SOM-RL ATSC also used the reduction of control delay as the operational objective to make the two approaches comparable to one another.

The intersection configuration modeled in VISSIM is shown in Fig. 5. Pedestrians and public transport in the form of the tram were excluded from the model for simplicity, but all signal programs followed all required safety precautions regarding

Table 2 FTSC signal program regime for intersection King Zvonimir street–Heinzelova street (Miletić 2019)

Time period	Signal program	Cycle duration [s]
06:00 AM–07:00 AM	2	90
07:00 AM–09:00 AM	3	100
09:00 AM–03:00 PM	2	90
03:00 PM–06:00 PM	4	100
06:00 PM–10:00 PM	2	90
10:00 PM–06:00 AM	1	60

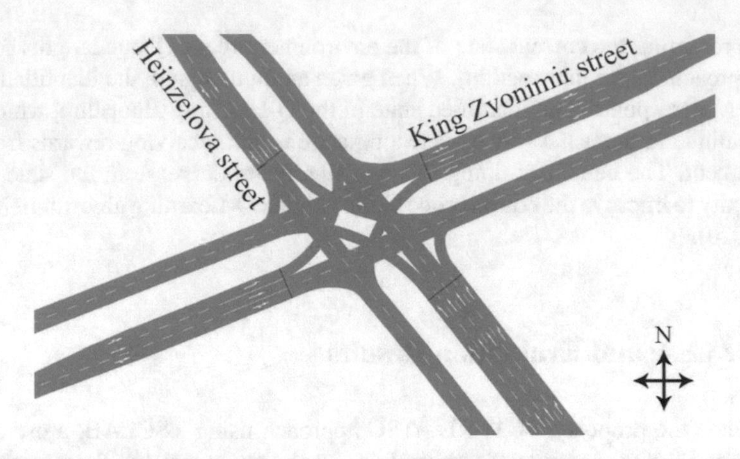

Fig. 5 Simulated intersection modeled in PTV VISSIM (Miletić et al. 2020)

the green light duration for pedestrians. The trams were excluded since they do not share road infrastructure with other vehicles. Due to data availability, the intersection was simulated for a regular workday from 06:00 AM to 10:00 PM.

5.2 *Obtained Simulation Results*

The obtained simulation results from TSCLab are presented here. Tables 3 and 4 show the average values of MoEs obtained per signal group for FTSC and SOM-RL approaches for the analyzed intersection. The cycle length for FTSC and SOM-RL approaches and traffic flow profile during the 16-h period of evaluations is displayed in Fig. 6. The vehicle arrivals per cycle and served vehicles during the green signal for both traffic signal control methods are shown in Fig. 7. Figures 8 and 9 show the green and red occupancy ratio per signal group for FTSC and SOM-RL traffic control approaches. Figure 10 presents the queue length at the end of each cycle for FTSC and SOM-RL traffic signal control approaches. The percentage of vehicle arrivals during the green signal for FTSC and SOM-RL traffic control approaches are presented in Fig. 11. Detailed analysis and interpretation of the simulation results follow in the next subsection.

5.3 *Evaluation of the Control Effects on Traffic Flows*

It can be noted in Fig. 6, that the change of the signal programs using the SOM-RL approach is more frequent compared to FTSC. Frequent signal program changes calculated online can cause significant transition delay and disruption to the coordinated operation of several consecutive intersections. Furthermore, the apparent

Table 3 Average values of MoEs for signal timing and throughput

Signal control	Signal group	Signal timing results			Throughput results		DI/LoS
		C	GT	Max GTUR	VApC	SVpGS	
FTSC	SG 1	91.37	25.47	0.85	16.12	13.34	38.5/D
	SG 2		15.74	0.87	8.01	3.49	
	SG 3		30.86	0.91	12.51	10.41	
	SG 4		17.30	0.79	5.53	3.90	
	SG 5		15.31	0.85	4.13	3.07	
	SG 6		20.70	0.86	3.09	2.24	
	SG 7		43.01	0.84	2.87	2.18	
	SG 8		21.34	0.69	5.25	4.56	
	SG 9		21.34	0.85	2.62	2.51	
Average values		91.37	23.45	0.84	6.68	5.08	D
SOM-RL	SG 1	64.78	16.39	0.55	11.33	8.50	33.8/C
	SG 2		13.41	0.74	5.65	2.20	
	SG 3		17.14	0.50	8.83	6.58	
	SG 4		13.72	0.62	3.92	2.46	
	SG 5		11.50	0.64	2.92	1.89	
	SG 6		12.19	0.51	2.17	1.39	
	SG 7		30.96	0.61	2.01	1.42	
	SG 8		14.33	0.46	3.88	3.33	
	SG 9		14.33	0.57	1.85	1.72	
Average values		64.78	16.00	0.58	4.73	3.28	C

Legend: C—cycle, GT—Green time, Max GTUR—Max green time utilization ratio, VApC—Vehicle arrivals per cycle, SVpGS—Served vehicles per green signal, DI—Control delay, LoS—Level of service

adjusting of signal duration to the short–term fluctuations in traffic demand can be seen. The SOM-RL approach has a shorter average cycle length than the fixed traffic signal control. The average cycle length duration is 64 s, which means that the signal program 1 with 60 s cycle length duration is the most used one during the simulation period. The average cycle length duration for the FTSC is 91 s, which means that signal programs 2 and 3 are the most used ones. Because of the longer average cycle duration, the number of vehicle arrivals per cycle, and the number of served vehicles during the green signal is higher in the case of FTSC (Fig. 7).

The Green Occupancy Ratio (GOR) is a measure used to describe the signal phase utilization, and it is an alternative measure for the saturation flow rate. Unlike GOR, the Red Occupancy Ratio (ROR) is a measure used to describe the efficiency of the signal phase, i.e., whether there remain unserved vehicles after the termination of the green signal and is determined for the first 5 s of the read signal. When GOR is 1 and ROR is 0, that means that the phase operates ideally, i.e., it is fully adapted

Table 4 Average values of MoEs for capacity and progression

Signal control	Signal group	Capacity results			Progression results		
		ROR_5	% VAoG	Max GTUR	VApC	SVpGS	AT
FTSC	SG 1	0.85	0.76	69.42	34.94	1.25	3.72
	SG 2	0.54	0.34	7.35	24.79	1.44	3.84
	SG 3	0.46	0.39	33.43	35.47	1.06	3.12
	SG 4	0.27	0.18	8.86	15.75	0.84	2.60
	SG 5	0.38	0.20	21.50	17.52	1.05	2.89
	SG 6	0.22	0.15	8.36	22.24	0.98	2.83
	SG 7	0.12	0.13	2.90	37.98	0.81	2.69
	SG 8	0.73	0.67	62.92	28.96	1.23	3.49
	SG 9	0.23	0.16	2.57	27.85	1.19	3.10
Average values		0.42	0.33	24.15	27.28	1.09	3.14
SOM-RL	SG 1	0.85	0.77	55.18	27.73	1.11	3.21
	SG 2	0.40	0.25	5.46	25.30	1.21	3.45
	SG 3	0.52	0.43	28.26	26.16	1.01	2.93
	SG 4	0.23	0.14	3.57	19.55	0.91	2.79
	SG 5	0.31	0.20	14.51	16.63	0.93	2.55
	SG 6	0.21	0.14	6.52	15.95	0.85	2.34
	SG 7	0.11	0.11	2.33	38.21	0.79	2.75
	SG 8	0.65	0.51	43.53	16.26	0.72	2.46
	SG 9	0.20	0.13	2.56	18.00	0.82	2.43
Average values		0.39	0.30	17.99	22.64	0.93	2.77

Legend: GOR—Green occupancy ratio, ROR5—Red occupancy ratio for the first 5 s, QLen—Queue length at the end of the cycle, % VAoG—Percentage of vehicle arrivals on green signal, VPR—Vehicle platoon ratio, AT—Arrival type

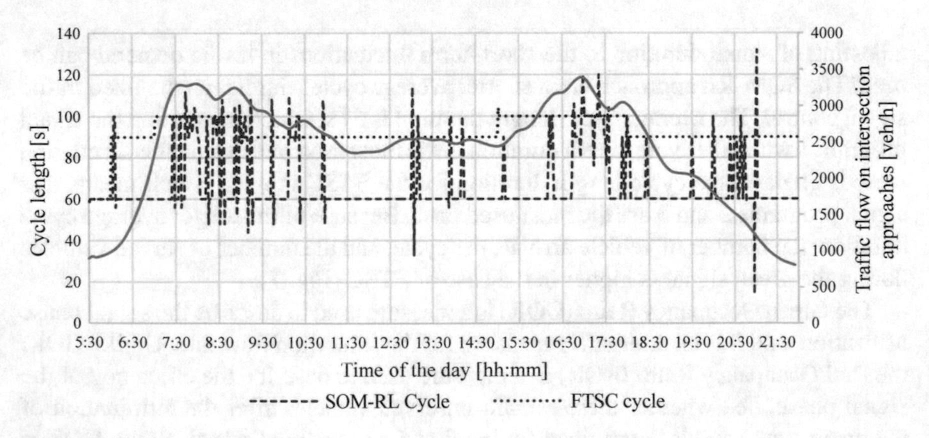

Fig. 6 Cycle length and traffic flow profile

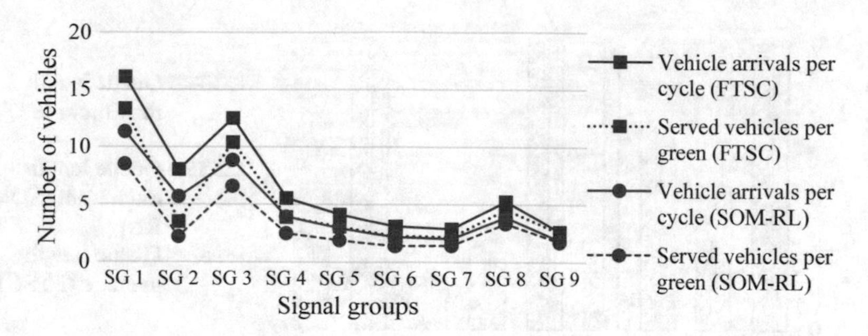

Fig. 7 Vehicle arrivals per cycle

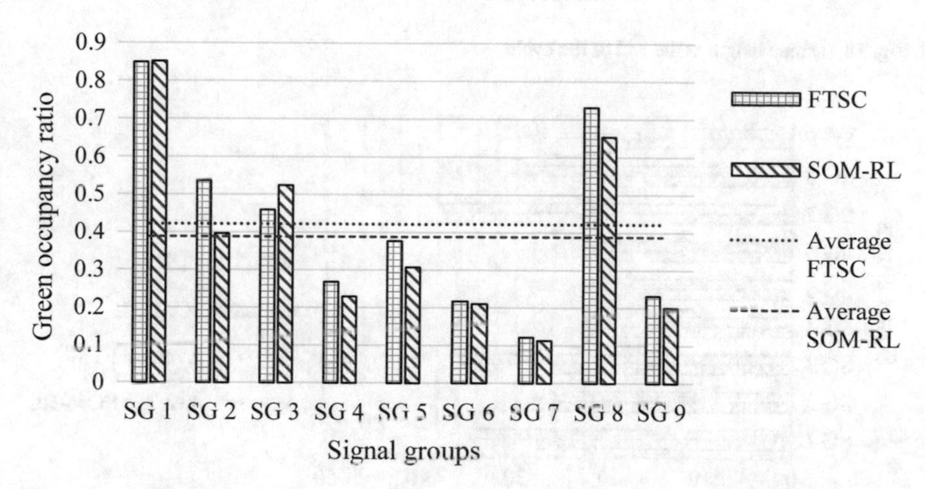

Fig. 8 Green occupancy ratio

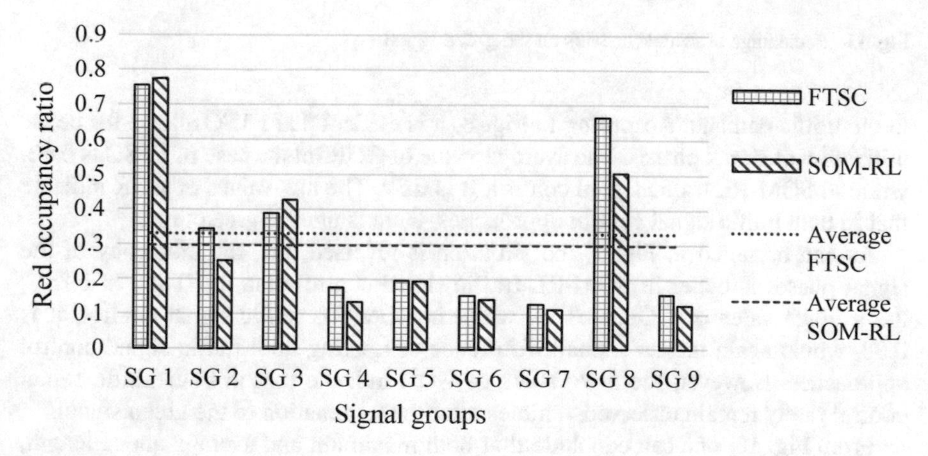

Fig. 9 Red occupancy ratio

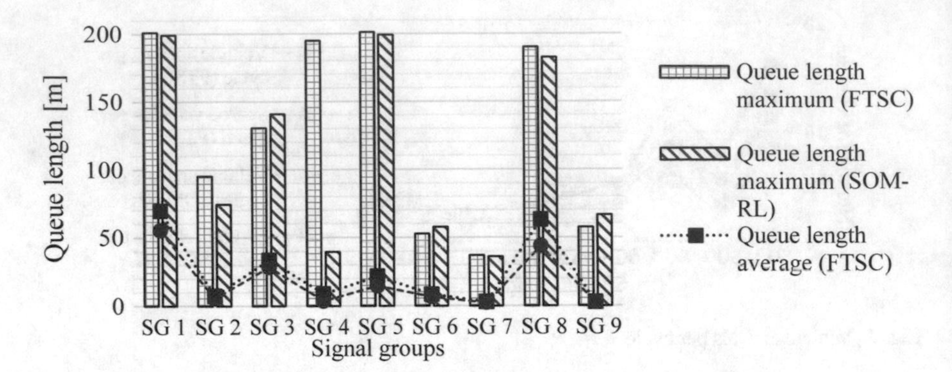

Fig. 10 Queue length at the end of the cycle

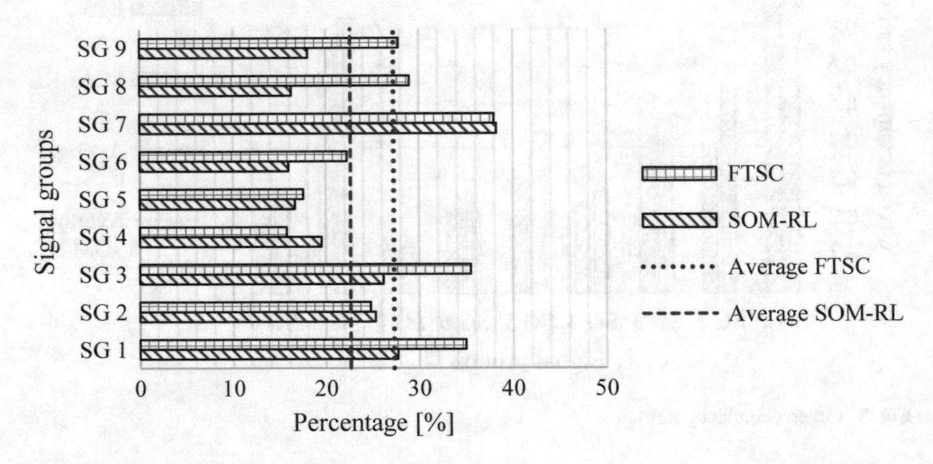

Fig. 11 Percentage of vehicle arrivals on the green signal

to the traffic demand. According to Fig. 8, it is evident that FTSC allows for better utilization of signal phases. The average value of GOR int the case of FTSC is 0.42, while in SOM-RL traffic signal control, it is 0.39. The low values of GOR indicate that in both traffic signal control approaches, there is unused green time.

As can be noted in Fig. 9, the situation is reversed, i.e., the efficiency of the signal phases is better in SOM-RL traffic signal control than in FTSC. In FTSC, the average value of GOR is 0.33, while in SOM-RL traffic signal control, it is 0.30, which again means a small difference comparing both traffic signal control approaches. However, the low values of GOR indicate that in both traffic signal control rarely remain unserved vehicles after the termination of the green signal.

From Fig. 10, one can conclude that both maximum and average queue length, which are measured at the end of each cycle, are lower in the SOM-RL approach compared to FTSC. In FTSC, the average queue length is 24.15 m, while in SOM-RL traffic signal control, it is 17.99 m which means 34% shorter average queue lengths at the end of each cycle for SOM-RL traffic signal control approach. This indicates that SOM-RL traffic signal control is more effective in handling traffic demand than FTSC.

The measure "Percentage of vehicle arrivals during the green signal" determines the percentage of vehicles arriving during the green signal. This measure is most commonly used to describe the quality of traffic flow progression between two adjacent intersections. Although this measure is not relevant for a single intersection, as shown in Fig. 11, the average percentage of vehicle arrivals during the green signal is higher in FTSC compared to SOM-RL traffic signal control approach. This is a consequence of the longer average cycle duration used by the FTSC approach.

The results show that the LoS with current FTSC used is classified as "D", while with the SOM-RL ATSC is classified as "C". This proves that there exist a significant reduction in overall control delay when the newly developed ATSC approach was applied.

6 Conclusion and Future Work

Simulation results show that the proposed SOM-RL traffic signal control offers significant benefits regarding traffic performances and demonstrate its superiority due to the capability to adapt to the traffic flow changes. The results show that SOM-RL based ATSC uses shorter cycle duration, achieving a lower green time utilization with the benefit of reducing queue lengths and thus improving the intersection throughput when compared to the existing FTSC. Analysis of cycle durations shows frequent cycle duration oscillations for SOM-RL based control during periods with increased traffic flow. While such frequent changes can provide an increase in performance, it should be noted that such frequent oscillations are not desired as they could lead to driver confusion. Additionally, it would be challenging to coordinate the controlled intersection with the rest of the network if the cycle duration is changed too frequently.

Future work on SOM-RL based ATSC will include a network of connected intersections for a more in-depth analysis of the impact of SOM-RL based ATSC applied on the entire managed traffic network. Further development of TSCLab will enable the analysis of larger traffic networks with multiple intersections under different traffic scenarios and traffic load conditions.

Acknowledgements The research presented in this paper is supported by the University of Zagreb Program Funds Support for scientific and artistic research (2020) through the project: "Innovative models and control strategies for intelligent mobility" and by the European Regional Development Fund under the grant KK.01.1.1.01.0009 (DATACROSS). The authors thank the companies "PTV Group" for providing a research license of the simulator VISSIM, and "Peek Promet" for providing the signal program samples. This research has also been carried out within the activities of the Centre of Research Excellence for Data Science and Cooperative Systems supported by the Ministry of Science and Education of the Republic of Croatia. The author Daniela Koltovska Nečoska received COST, CEEPUS and ERASMUS+, and author Edouard Ivanjko received ERASMUS+ mobility grants that enabled the research presented in this paper.

References

Dakic I et al (2018) Upgrade evaluation of traffic signal assets: high-resolution performance measurement framework. PROMET Traffic Transport 30(3):323–332

Day C et al (2014) Performance measures for traffic signal systems: an outcome-oriented approach. Purdue University, West Lafayette, Indiana, USA

Day C et al (2015) Integrating traffic signal performance measures into agency business processes. Purdue University, West Lafayette, Indiana, USA

El-Tantawy S et al (2013) Multiagent reinforcement learning for integrated network of adaptive traffic signal controllers (MARLIN-ATSC): methodology and large-scale application on downtown Toronto. IEEE Trans ITS 14(3):1140–1150

Gettman D et al (2013) Measures of effectiveness and validation guidance for adaptive signal control technologies. US Department of Transportation, Federal Highway Administration

Mannion P et al (2016) An experimental review of reinforcement learning algorithms for adaptive traffic signal control. In: McCluskey T, Kotsialos A, Müller J, Klügl F, Rana O, Schumann R (eds) Autonomic road transport support systems. Autonomic Systems. Birkhäuser

Michailidis IT et al (2018) Autonomous self-regulating intersections in large-scale urban traffic networks: a Chania city case study. In: 2018 5th international conference on control, decision and information technologies (CoDIT), pp 853–858

Miletić M (2019) Adaptive control of isolated signalized intersection using neural networks. Master thesis. Faculty of Transport and Traffic Sciences, University of Zagreb, Croatia

Miletić M et al (2020) State complexity reduction in reinforcement learning based adaptive traffic signal control. In: 62nd international symposium ELMAR-2020, Zadar, Croatia

Pavleski D et al (2020) Development of TSCLab: a tool for evaluation of the effectiveness of adaptive traffic control systems. In: Karabegović I (ed) New technologies, development and application II. NT 2019. Lecture notes in networks and systems, vol 76. Springer, Berlin

Touhbi S et al (2017) Adaptive traffic signal control: exploring reward definition for reinforcement learning. Proc Cinozter Sci 109:513–520

Vujić M (2013) Dynamic priority systems for public transport in urban automatic traffic control. Dissertation. Faculty of Transport and Traffic Sciences, University of Zagreb, Croatia

Wahlstedt J (2013) Evaluation of the two self-optimising traffic signal systems UTOPIA/SPOT and ImFlow, and comparison with existing signal control in Stockholm, Sweden. In: Proceedings of the 16th international IEEE annual conference on intelligent transportation systems (ITSC 2013), pp 1541–1546

Discrete Simulation Model for Urban Passenger Terminals

Aura Rusca, Eugen Rosca, Florin Rusca, Mihaela Popa, Cristina Oprea, and Oana Dinu

Abstract The recent emergence of bad events (fires, terrorist attacks, etc.) resulted in a large number of deceased and injured people, brought to the attention of specialist's problems and caused by the lack of adequate dimensioning of access and transit path. In the case of urban transport terminals linking rail lines, metro lines, and surface public transport lines there are a number of specific aspects such as the transport means arrival after a fixed schedule, group acces within the terminal, long waiting time in the terminal etc. In this case the use of a discrete simulation model allows the evaluation of the use of access gates, transit times, etc. In this paper, a simulation model is developed for a hypotetical passenger terminal using the topology of the main Romanian train station. The obtained data can be used for the optimal dimensioning of the number of access gates, of the stairs, of the waiting area etc.

Keywords Passenger terminals · Discreet simulation · Passenger flows

---

A. Rusca (✉) · E. Rosca · F. Rusca · M. Popa · C. Oprea · O. Dinu
Faculty of Transport, University Politehnica From Bucharest, Spl. Independentei, No 313, District 6, Bucharest, Romania
e-mail: aura.rusca@upb.ro

E. Rosca
e-mail: eugen.rosca@gmail.com

F. Rusca
e-mail: florin.rusca@upb.ro

M. Popa
e-mail: mihaela.popa@upb.ro

C. Oprea
e-mail: cristina.oprea@upb.ro

O. Dinu
e-mail: oana.dinu@upb.ro

M. Petrović and L. Novačko (eds.), *Transformation of Transportation*, EcoProduction,
https://doi.org/10.1007/978-3-030-66464-0_5

1 Introduction

In the last decade, a series of events with unfavorable effects for the public have attracted the attention of decision-makers on the importance of the correct dimensioning of access ways. Thus, the outbreak of fires in some clubs or event rooms led to an overloading of the access gates with negative consequences quantified in deceased or injured people. Also, the increase of terrorist attacks risk puts pressure on the building owners to dimension the access gates so that they allow the passing of a large flow of people.

In the case of passenger transport terminals, several specific issues need to be considered. Thus, the arrival of the passengers is mainly in groups when the means of transport arrive in station (train, subway, tramway, etc.). This leads to a temporary increase of the loading degree of the access gates and transit path used for the transfer between different means of transport. For the situation when we have an urban passenger terminal that connects the underground and surface urban transport with the railway transport, it is necessary to consider the dimensioning of a waiting area for the passengers.

The transition between the underground and the surface network is necessary to be realised by using the stairs that can be simple stairs or escalators. Their transit capacity is extremely important in designing the terminal. The capacity of the platform for the underground transport must allow both transit and stationing of the passengers.

2 Literature Review

The study of the topological aspects of passenger terminals is of particular interest to the transports specialists. Thus, in the literature, there are a number of papers that treat the passenger terminal both in terms of transit capacity, terminal design, estimation of passenger flows and safety and security level of the passengers.

In their paper Nommika and Antov (2017) are studying passenger terminal capacity assessment models at airports and propose a new model for estimating the capacity of the terminal in relation to the dynamics of the passengers' flow (Nõmmika and Antov 2017). Also, in their work Solak and all (2009), it is developed a stochastic multistage optimization model used in assessing the capacity of a new passenger terminal (Solak et al. 2009). The holistic determination of the terminal's capacity allows to overcome the uncertainty regarding the moment of occurrence of the traffic peak, respectively of the maximum flow value in this this peak moment.

The evaluation of the passenger risk and safety at the terminal is realised in their papers by Gromulea and all (2017) and Yatskiv and all (2016). Thus, in the first paper, there are developed a series of rules and norms that need to be followed in order to ensure the safety of the passengers and to improve their security (Gromulea et al. 2017). By using a simulation model in SimWalk, in the second paper, critical areas in the terminal are identified for which measures are required not to affect the safety and security of passengers (Yatskiv et al. 2016). Transit through a security checkpoint inside the terminal is analysed in the paper of Popa and Strer (2016) in terms of exploitation costs. There are identified methods of reducing these costs without leading to a degradation of the security level in the terminal (Popa and Strer 2016).

Basic parameters for the design of intermodal public transport are presented in the paper of Margarita and all (2016). A series of eight passenger terminals in which two or more modes of transport converge in terms of intermodal integration and innovative design attributes from each terminal are analysed (Margarita and Durán 2016). Also, in the paper of Pitsiava-Latinopouloua and Iordanopoulosb (2012), aspects of passenger terminal design, in terms of design standards, are found to provide a better level of passenger service (Pitsiava-Latinopouloua and Iordanopoulosb 2012).

Estimates of passenger flows within the passenger terminal can be found in the papers of Ahn and all (2017), Liu and Chen (2017), Rusca and all (2013). The first paper evaluates the passenger flow in the terminal using a classical four-step model (Ahn et al. 2017). The results obtained are used to identify the critical points in the terminal for which passenger traffic microsimulation models are developed. The second paper presents an evolutionary algorithm of the type of neural networks used in the estimation of the passenger flows (Liu and Chen 2017). Finally, in the last paper the authors evaluate the degree of loading of the ticket offices in relation to the passenger train timetable arriving/leaving in a passenger terminal (Rusca et al. 2013).

From the point of view of the method used in the evaluation of the passenger terminals, the use of simulation models occupies an important place. These are realised using software dedicated to pedestrian flow modelling (Yatskiv et al. 2016; Pitsiava-Latinopouloua and Iordanopoulosb 2012; Ahn et al. 2017).

A special case is the use of software for discrete activity simulation inside the passenger terminal (Rusca et al. 2013; Fonseca et al. 2014). This allows the determination and the evaluation of a set of service parameters perceived by the passengers within the terminal. The development of a discrete simulation model presents a certain flexibility that can be adapted to the case of an urban intermodal terminal in which several surface or underground, respectively regional and interregional transport modes converge.

3 Passenger Terminal Safety Design Aspects

Transport terminals represent junction points between local, regional and interregional transport. The main objective of a terminal is to serve transport demand by maintaining the travellers' safety and comfort. The main users of a public transport terminal are the travellers. When a terminal is designed all the factors that contribute to the size of the flows into the terminal must be taken into consideration. Thus, a pedestrian's flow in a terminal is formed of both passengers and their attendants. For the first ones, the terminal represents:

Table 1 Activities in a transport terminal

Activity	Actors	Facilities
Access, entrance	Travellers, attendants, employees	Doors, gates, tourniquets
Information	Travellers, attendants, employees	Bgoards, electronic displays, signs, arrows
Movement	Travellers, attendants, employees	Corridors (walkways), stairs, escalators, lifts or moving walkways
Waiting	Travellers, attendants	Waiting areas, platforms, queues
Buying tickets	Travellers	Ticket offices, ticket machines
Shopping	Travellers, attendants	Commercial areas (shops)
Exit	Travellers, attendants, employees	Doors, gates

- the origin station from which they start their journey.
- the destination station where the journey ends.
- the transfer station between different modes of transport (surface or underground, urban or regional, etc.).

In the case of the attendants, they either travel together with the travellers when they leave or wait for them to arrive. A special category is represented by the employees inside the terminal (cleaners, employees from the shopping area, etc.) but which, by their small number, do not influence the terminal's capacity.

The different areas of the terminal are linked through corridors (walkways), stairs, escalators, lifts or moving walkways.

In Table 1 the activities from a transport terminal, the actors that execute them (the users) and the needed facilities are centralised.

The transport operator must maintain a high level of service inside the terminals reflected through the fluidity of travellers' flow and by inducing a sense of concern for their safety, convenience and comfort. A pleasant waiting environment must be created focusing on the users' psychological and physiological aspects and also on maintaining the service discipline, as follow:

- *Psychological aspects*:
  The most important cause of stress in our days is induced by the lack of control (Bateson 1985). Because the waiting time in a queue cannot be controlled by users, it represents a very unpleasant experience for them. By giving them the control and by offering them activities that they can do while waiting (something to read, to eat or to watch), information regarding the estimated waiting time, providing FIFO serving discipline avoiding queue re-ordering, their stress can be diminished. A proper social behaviour is the key to a waiting tolerated by users.
- *Physiological aspects*:
  The waiting space can contribute to the traveller's mood. The waiting area should have a pleasant architecture and should be nicely decorated (Maister 1985). Furthermore, this should be designed so that:

- noise—the accepted noise level in a waiting area is 40–48 dB
- illumination—the illumination in a big room should be irregular for relaxation, while in a small room it should be uniform to create the space impression
- climate and ventilation—for the travellers that wait the indicated temperature in winter is between 20° and 24 °C and in summer between 22° and 27 °C; the humidity in summer should be between 20 and 30% and in winter between 50 and 80% are adequate.

• *Maintaining the service discipline*:
Travelers should be compelled to respect the service discipline. The implementation of a service discipline requires certain effective means for tracking and selecting travellers. These means can be passive or active. Passive constraints require the provision of information, verbally or non-verbally, which leads to queuing. Active constraints directly involve serving points in the selection and ordering of passengers.

Some aspects of designing a service system in the public transport terminals are emphasized in Table 2.

In accordance with the national legislation related to buildings, rooms and other built-up areas, evacuation routes should be provided to allow people to reach out to the ground at the shortest possible time and under safe conditions (Romanian legislation). Determining the size of the evacuation routes (for more than five people) is to determine the required gauge and to determine the length of the routes to ensure rapid evacuation from the building. The number of passage units (flows) to be provided for the evacuation of persons and the gauges required for the passage of the evacuation flows shall be take in accordance with national legislation.

Table 2 Aspects of designing a service system in a transport terminal [after (Rosca et al. 2004))]

Psychological aspects	Prompt information	– Estimating the waiting time; – Quick notification of irregularities; – Permanent contact between the system and the service demands
	Equity	– Using FIFO (First In–First Out); – Eliminating unjustified priorities; – Allocating different service stations for demands with different priority levels;
Physical aspects	Noise	– Powerful noise removal using soundproofing enclosures and noise absorbing materials
	Illumination and ventilation	– Ensuring adequate lighting, both natural and artificial; – Installation of ventilation equipment
	Congestion	– Allocating sufficient space for waiting; – Installation of protective barriers

4 The Simulation Model for Passengers Terminal

To develop simulation models adapted to the passenger terminals activity, a series of dedicated modelling software is used. Some of these are developed by manufacturers to simulate the activity in transport terminals, respectively for creating discrete simulations.

From the point of view of how the simulation model is developed, the simulation software can be performed (Academies and of Sciences, Engineering, and Medicine. 2010):

- continuous simulation model:
  - PaxSim
  - EDS-SIM
  - TRACS
  - Total AirportSim
  - Baggage Systems
- discreet simulation model:
  - SIMULATION MODEL DESIGN
  - ARENA Simulation Rockwell
  - SIMUL8
  - VISWalk.

In the case of discrete event simulation models, they can be used within a wider spectrum, without, however, affecting the quality of the modelling. They also allow a greater flexibility and adaptability in the construction of new modelling for passenger terminals.

The purpose of these modelling is to provide a support for passenger terminal planners and designers. They allow the testing and validation of the proposed solutions, as well as the assessment of the passenger terminal's adaptability to the risk limit situations. Critical areas where passengers' safety and security can suffer can be identified.

When the flow of passengers that enter the terminal has several specific features such as:

- it has a stochastic character
- the number of passengers that arrive in terminal follow different distribution laws
- in the case of transport means, the arrival of the passengers is in groups
- the load of access ways, stairways or elevators is uneven in time
- etc.

Using a discrete simulation model allows the terminal administration to identify critical areas, to dimension properly the new facilities within the terminal, to identify the size of the required staff, etc.

In the paper we considered the case of a fictitious terminal (Fig. 1) with features like the North Railway Station intermodal terminal in Bucharest. The arrival of

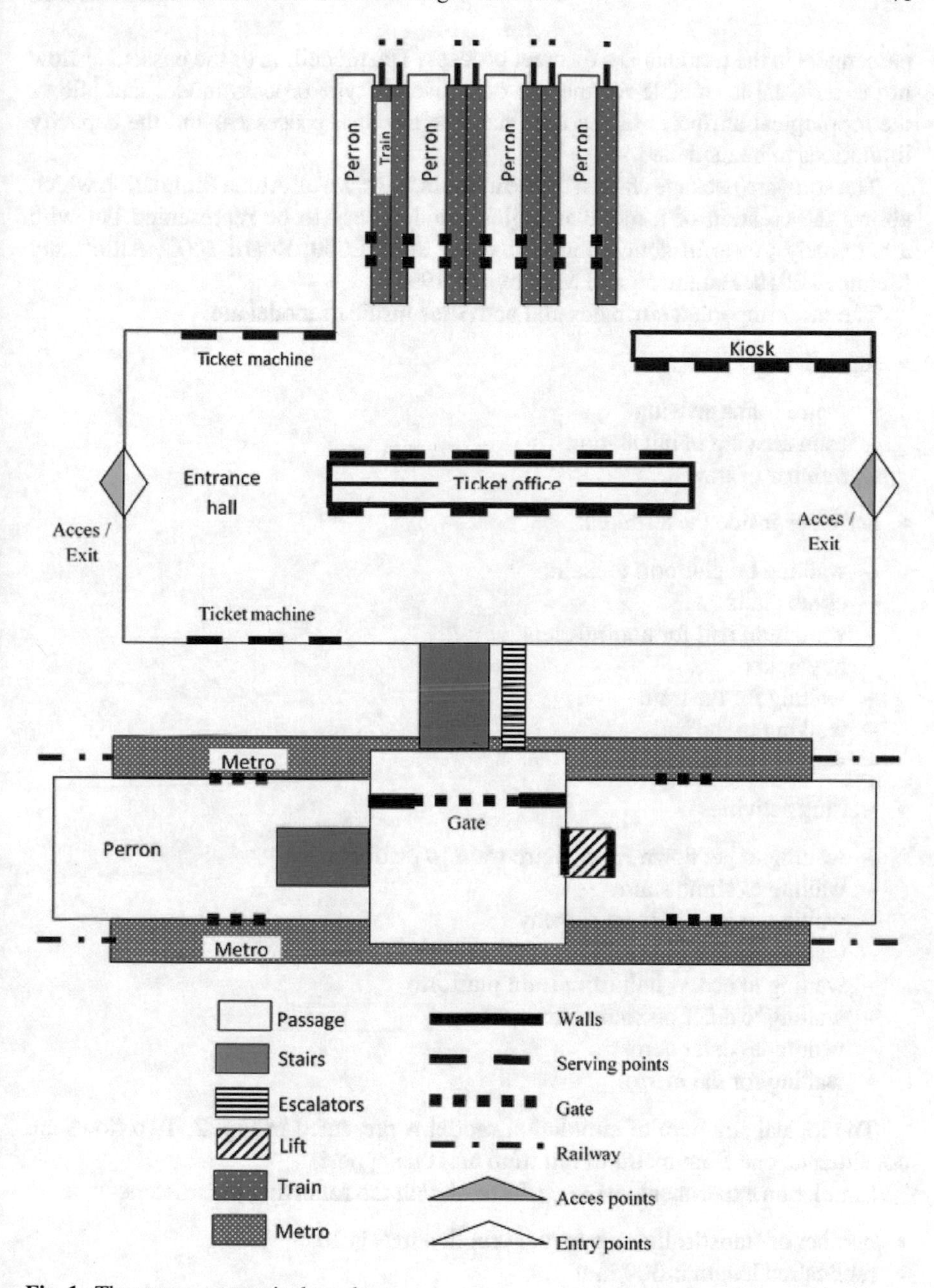

Fig. 1 The passenger terminal topology

passengers in the terminal is a discreet process. The modelling of the passenger flow in the terminal is suitable for the use of a discrete type process model that allows the topological attributes of the terminal, the terminal processes, and the capacity limitations to be modelled.

The software package chosen by the authors is Rockwell Arena Simulation which allows the creation of a modular logical model, easy to be represented but with a extremely powerful simulation tool (Law et al. 2000; Board 2005; Altiok and Melamed 2010; Hammann and Markovitch 1995).

The most important attributes and activities inside of model are:

- input data
  - metro trains arriving
  - train arriving in rail station
  - number of travellers
- activities inside the terminal:
  - walking on platform to stairs
  - climb stairs
  - waiting in hall for nontravelers
  - buy ticket
  - waiting for the train
  - walking to the hall
  - walking to the metro
- waiting activities
  - waiting to get down from metro train to platform
  - waiting to climb stairs
  - waiting to get out from subway
  - waiting for the train
  - waiting to access hall from train platform
  - waiting to enter on stairs from metro
  - waiting to enter at metro
  - waiting for the metro.

The logical structure of simulation model is presented in Fig. 2. Two flows are considerate, one from metro to rail train and one opposite.

Simulation experiments were performed with the following parameters:

- number of statistically independent replications is 10
- replication length is 300 min
- warm-up period is 60 min
- different scenarios for train inter-arrivals distribution, dimension of storage zone, number of enter or exit gates from subway.

The symbols used in describing scenarios are:

- trains inter-arrival times to station, TT

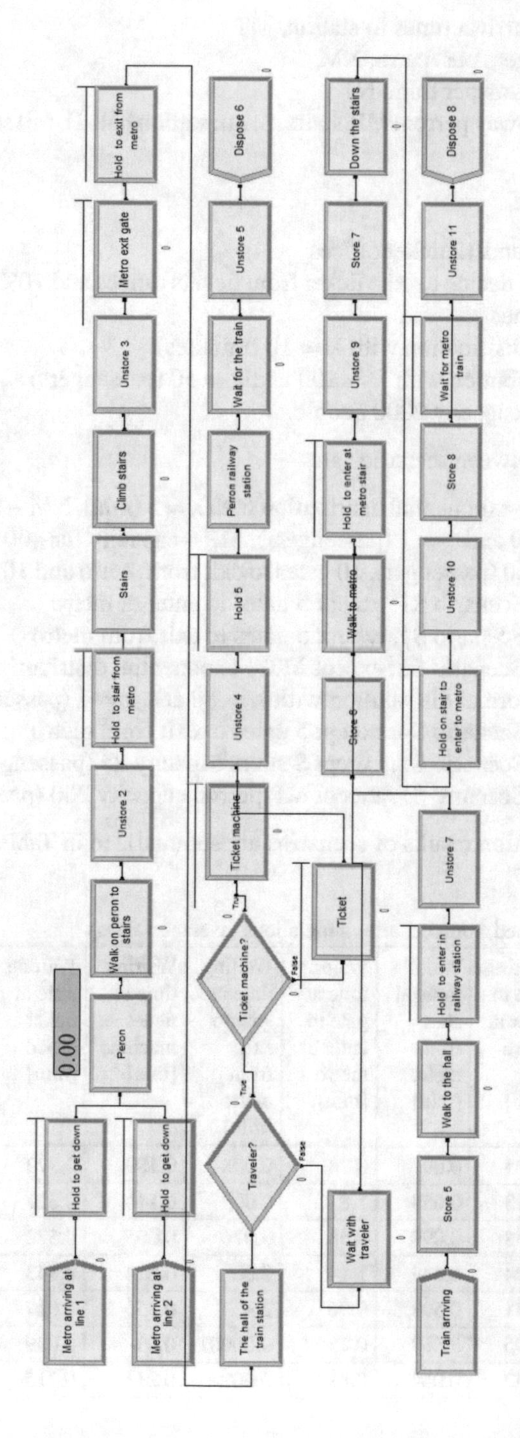

Fig. 2 The logical structure of simulation model

- metro trains inter-arrival times to station, MT
- number of passengers per metro, NM
- number of passengers per train, NT
- storage size for subway perron, MP, stairs, S, rail station hall-H, rail station perron-RP.

All scenarios have:

- 6 ticket machines and 12 ticket offices
- 90% of passengers decide to take ticket from tickets offices and 10% of passenger from ticket machines
- TT—exponential distribution with $\lambda = 10$ (minutes)
- NT—normal distribution with $\lambda = 200$ and $\sigma = 50$ (passengers)
- H-rail station hall capacity 2000 people.

The differences between scenarios are:

- Scenario S1: MT—exponential distribution with $\lambda = 5$ (min), NM—normal distribution with $\lambda = 40$ and $\sigma = 1$ (passengers), MP—capacity for 400 (passengers), S—capacity for 400 (passenger), 10 gates to exit from metro and 10 gates to enter
- Scenario S2: like Scenario S1, except 5 gates to enter at metro
- Scenario S3: like Scenario S1, except 5 gates to exit from metro
- Scenario S4: like Scenario S1, except MT—exponential distribution with $\lambda = 8$ (min) and NM—normal distribution with $\lambda = 80$ and $\sigma = 1$ (passengers)
- Scenario S5: like Scenario S4, except 5 gates to exit from metro
- Scenario S6: like Scenario S1, except S stairs capacity 25 (passengers)
- Scenario S7: like Scenario S1, except MP perron capacity 200 (passengers).

The output simulation results of scenarios are summarized in Table 3.

Table 3 Statistics obtained from scenarios simulations (average values)

	Entity number [passengers]	Waiting time to enter at metro stair [min]	Waiting time on stair from metro [min]	Waiting time at gate to enter at metro [min]	Waiting time at gate to exit from metro [min]	Waiting time at ticket machine [min]	Waiting time at ticket office [min]	Length of queue to enter at metro stair [passengers]
S1	7490	0.035	0.09	0.78	0.003	0.189	0.772	0.884
S2	8587	0.013	0.053	1.8	0.003	0.144	0.627	0.855
S3	8688	0.058	0.099	0.68	0.576	0.095	0.572	1.657
S4	9890	0.424	0.689	1.14	0.006	0.824	2.843	14.2
S5	9219	0.331	0.598	0.96	2.237	0.353	2.017	9.277
S6	8414	4.695	3.867	0.75	0.00001	0.03	0.039	134.38
S7	8375	0.042	0.064	0.81	0.002	0.211	0.715	1.135

Scenario 4 and scenario 5, with inter-arrival times exponential distribution with $\lambda = 8$ min and number of passenger from metro train normal distribution with $\lambda = 80$ and $\sigma = 1$, has the number entity in simulation greater than all simulations scenarios (as shown in Table 3). But, the average waiting time to enter at metro stair it is also higher than in scenarios 1, 2, 3 and 7. However, it can be seen that scenario 6, where the stair capacity decreases by half, the average waiting time increases over 10 times. In this case, the staircase transit capacity presents a vulnerability that needs to be carefully monitored. The negative quantifiable effects are in this case the increase in waiting times to enter or exit the subway and queue length, respectively. The influence of reduction of exit or entrance gates can be estimated but it is relatively small.

5 Conclusions

Computers simulation models allow the evaluation of the use of access gates, transit times, etc. and the identification of the vulnerable areas. Also, the negative impact caused by these vulnerable areas can be assessed and ways to reduce it can be identified. In addition the data obtained can be used for the optimal dimensioning of the number of access gates, of the stairs, of the waiting areas etc. The obtained results prove the idea that modelling of the passenger flow in the terminal is suitable for the use of a discrete type process model. In this can be introduced the topological attributes of the terminal, the terminal processes and the capacity limitations to be modelled. In our paper presented case the logical model have two flows for passengers. One is from underground transport system to railways trains and the second is from railway station to metro train. The input data considered are: metro trains/railway trains inter-arrival times distribution, number of passenger, transit capacity for metro/railway perron, enter/exit gate, serving time distribution at ticket machines and ticket office, etc. The results obtained by simulation identify a problem in the metro access stairs whose capacity variation leads to a considerable degradation of the service parameters. For these, the administrator of the passenger terminal needs to be more careful as it is a critical area of the terminal. The research was based on real topological aspects but for which mathematically generated data were used. Our future goal is to use the model developed with data collected from major passenger terminals. We have the possibility to qualitately test the results obtained and to calibrate the model correctly.

Acknowledgements This work has been funded by the European Social Fund from the Sectoral Operational Programme Human Capital 2014-2020, through the Financial Agreement with the title "Scholarships for entrepreneurial education among doctoral students and postdoctoral researchers (Be Antreprenor!)", Contract no. 51680/09.07.2019 POCU/380/6/13 - SMIS code: 124539.

References

Ahn Y, Kowada T, Tsukaguchi H, Vandebona U (2017) Estimation of passenger flow for planning and management of railway stations. Transport Res Proc 25:315–330

Altiok T, Melamed B (2010) Simulation modeling and analysis with Arena. Elsevier

Bateson JEG (1985) Perceived control and the service encounter. In: Czepiel JA, Solomon MR, Surprenant CF (eds) The service encounter, managing employee/customer interaction in service businesses. Lexington Books, Lexington, Mass

Board A (2005) Stochastic modelling and applied probability. Springer, Berlin

Fonseca P, Casas I, Casanovas J, Ferran X (2014) Passengers flow simulation in a hub airport: an app. to the Barcelona International Airport. Simul Modell Pract Theor 44:78–94

Gromulea V, Yatskiv (Jackiva) I, Pēpulisa J (2017) Safety and security of passenger terminal: the case study of riga international coach terminal. Proc Eng 178:147–154

Hammann JE, Markovitch NA (1995) Introduction to Arena [simulation software]. In: Winter simulation conference proceedings

Law AM, Kelton WD, Kelton WD (2000) Simulation modeling and analysis. McGraw-Hill, New York

Liu L, Chen R (2017) A novel passenger flow prediction model using deep learning methods. Transp Res Part C 84:74–81

Maister DH (1985) The psychology of waiting lines. In: Czepiel JA, Solomon MR, Surprenant CF (eds) The service encounter, managing employee/customer interaction in service businesses. Lexington Books, Lexington, Mass

Margarita L, Durán BM (2016) Basic parameters for the design of intermodal public transport infrastructures. Transport Res Proc 14:499–508

Nõmmika A, Antov D (2017) Modelling regional airport terminal capacity. Proc Eng 178:427–434

Pitsiava-Latinopouloua M, Iordanopoulosb P (2012) Intermodal passengers terminals: design standards for better level of service. Proc Soc Behav Sci 48:3297–3306

Popa A, Strer J (2016) Analysis of passenger and vehicle flows with microscopic simulations as a result of security checks at ferry terminals. Transport Res Proc 14:1384–1393

National Academies of Sciences, Engineering, and Medicine (2010) Airport passenger terminal planning and design, volume 2: spreadsheet models and user's guide. The N. A. Press, Washington, DC

Rosca E, Rusca F, Panica A (2004) Aspecte ale calităţii servirii călătorilor în terminalele de transport (romanian), Buletinul AGIR, An IX, nr.3

Rusca F, Rosca E, Rosca M, Rusca A (2013) Queueing systems with non-uniform arrivals applied to estimate services quality in passengers transport terminals. In: Advances in automatic control, modelling & simulation—proceedings of the 15th international conference on automatic control, modelling & simulation, recent advances in electrical engineering series, vol 13, pp 86–91

Solak S, Clarke J, Johnson E (2009) Airport terminal capacity planning. Transp Res Part B 43:659–676

Yatskiv (Jackiva) I, Savrasovs M, Gromule V, Zemljanikins V (2016) Passenger terminal safety: simulation modelling as decision support tool. Proc Eng 134:459–468

Characteristics of Departing Passenger Reports to the Passport Control Queuing System

Mateusz Zając and Maria Pawlak

Abstract The aim of the article was to present the results of the research, which concerned the analysis of the stream of passenger reports to the queuing system of passport control in the departure hall. These studies are very useful and may be used in the future by those stakeholders who want to develop a simulation model for such a system. This work fills a gap in the scientific literature. There are many different works that are related to security control or check-in. Passport control has so far been neglected. Particular interest should be shown by those who manage a system where passport control is directly in front of the departure gates. This structure of the system may then be a bottleneck in the process of departing passengers handling. This article provides details of the various air operations within many destinations.

Keywords Passenger service · Airport terminal · Passport control

1 Introduction

For several years now, there has been a dynamic increase in air traffic causing technical and organizational problems related to its handling in the air and on the ground. This is a challenge in terms of guaranteeing punctuality in air operations. Studies carried out by Neufville (2008) signal the required increase in capacity of Pair transport infrastructure. This is due to the continuous development of air transport. The covid-19 epidemic caused a sudden collapse in air traffic. This traffic is systematically restored, but there will be some time before the number of passengers will fully be recovered. This time is an opportunity to adapt all passenger service systems to the return of this heavy workload. After all, so far, there have been very long delays,

M. Zając (✉) · M. Pawlak
Wroclaw University of Science and Technology, Wyb. Wyspianskiego 27, 50-370 Wroclaw, Poland
e-mail: mateusz.zajac@pwr.edu.pl

M. Pawlak
e-mail: maria.pawlak@pwr.edu.pl

M. Petrović and L. Novačko (eds.), *Transformation of Transportation*, EcoProduction,
https://doi.org/10.1007/978-3-030-66464-0_6

which resulted from the system overload (CODA Digest All-causes delay and cancellations to air transport in Europe, Report 2019). This can be avoided because now there is an opportunity to change the structures of processes and learn how to manage them better, when there is a moment of breathing. It is therefore important to know the actual flow characteristics of passenger movements—from just before the outbreak.

The practice of modelling logistics systems requires the degree of complexity of the system representation to be determined. These models can be developed on a macroscopic scale that reflects the little details of such a system. The opposite can be true for microscopic models that contain a lot of details. A common feature for airport passenger queuing systems is that each model requires a notification stream od passengers to be estimated. This is, of course, about models that are used to plan passenger service. The exception is a system performance analysis which examines the maximum capacity of service desks.

This article presents the results of research on the real passport control system in the departure hall of Wroclaw Airport. The aim of this work was firstly to provide real data so that it could be used in simulation models by other stakeholders. This can be a great convenience for those who have so far assumed the theoretical input data. The second aim of the article was to analyse whether these data depend on the type of flight destination. This objective will make it possible to check whether the use of simulation models for passport control on a micro or macroscopic scale is justified.

Chapter "Traffic Flow Simulators with Connected and Autonomous Vehicles: A Short Review" provides a literature review of how passenger handling systems at the airport have been modelled so far. The research gap was indicated there. Chapter "Application of Dimensionless Method to Estimate Traffic Delays at Stop-Controlled T-Intersections" describes the research assumptions and the actual system. It is very useful to know where the research results can be applied. After all, stakeholders must have a similar system at their airport to be able to rely on these results. The results are presented in Chap. "In-Depth Evaluation of Reinforcement Learning Based Adaptive Traffic Signal Control Using TSCLAB". A summary of the work is provided in Chap. "Discrete Simulation Model for Urban Passenger Terminals".

2 State of Art

There are three types of approaches to modelling passenger service at an airport in the scientific literature. The dedicated approach is to cover only one of the passenger service subsystems. However, some of the works carry out an analysis for several, occurring in succession, passenger service processes (semi-holistic approach). A more complex analysis is presented by works that take into account all processes carried out on the terminal (holistic approach).

The first deterministic models of airport passenger service systems were developed in a holistic approach (Newell 1971). In the paper (Brunetta et al. 1999), this approach found its practical application in the form of a computer application. The deterministic model, which is based on a mathematical model, is also presented in Hassan et al. (1989). The passenger service system was presented as a hierarchical structure and the author focused on checking the relationships between passenger service subsystems. The dependence of subsequent service subsystems was the subject of other scientific works. The authors of the paper (Solak et al. 2009) believe that their previous works, using a holistic approach, did not allow for appropriate analyses. Therefore, they proposed a stochastic model, which treats subsequent service subsystems as dependent processes. The analysis was carried out for the profiles of reports occurring during the hourly peak. The paper (Gatersleben and Wei 1999) presents a simulation model which, on the basis of 5-year studies on the real system, was to help to determine the critical path in the flow of passenger streams at the airport. The transition time between individual peaks of the graph was determined taking into account the number of passengers in individual terminal areas. The paper (Schultz and Fricke 2011) presents a model of passenger flow through an airport based on an agent approach. It is a stochastic model, in which the dynamics of passenger movement between subsequent service subsystems is mapped in detail. The development of the agent approach in passenger flow modelling is also presented in Ma et al. (2012). Here too, passengers are autonomous objects, making decisions in a dynamic way, depending on the environment.

In the semi-holistic approach, in the paper (Eilon and Mathewson 1973), a model of arriving passengers service was developed. The operation of the simulation model consisted in placing the passenger successively in particular service subsystems. Output data were estimated on the basis of the time of reporting and time of passenger service, using regression models. An alternative solution allowing for conducting a wider spectrum of analyses is presented in the paper (Kobza and Jacobson 1996). Part of the data is given by probability distributions matched on the basis of real data from the airport in New York. Part of the processes is described by deterministic times of transitions between particular service subsystems. The model was extended (Jim and Chang 1998) with additional graph tops covering additional services such as: restaurants, information points, etc. In the model, service time characteristics are implemented in particular subsystems. In the model, a direct transition between individual service subsystems was assumed. The stochastic microscopic model is presented in the paper (Guizzi et al. 2009). The model has a modular structure and is dedicated to check-in passengers departing from the airport. The paper focuses on a detailed mapping of the ticket and baggage check-in and security control systems. The work also paid attention to the impact of the schedule management of check-in desks on the flow of passenger applications to security control.

By far the largest number of scientific papers use a dedicated approach. The paper (Boekhold et al. 2014) presents a simulation model, which was used to analyze the sensitivity of the implementation of the screening process in the aspects of the impact of employee performance, initial screening, random alarm rate, the number of baggage held on the performance of the established screening system with a given stream of reports. In the paper (Leone 2002), on the basis of the conducted simulation experiments, the relation between the number of triggered alarms and the efficiency of the security control area was also determined. In the paper (Leone and Liu 2003), the actual efficiency of the hand luggage control process after the introduction of regulations requiring ETD control was determined. The results were compared with the values of theoretical efficiency of the device. In the work (Kierzkowski and Kisiel 2015a), the flow of passengers was modelled to compare two variants of configuration of security control stations—single and double. Then, in the study Kierzkowski and Kisiel (2015b) the simulation model was used to analyze the sensitivity of the size of passenger preparation and baggage reception areas to the performance of the screening station for a double station and in the study Kierzkowski and Kisiel (2015c) for a single station. The paper (Li et al. 2018) presents the application of computer simulation to compare six different passenger queuing structures before screening. The authors of the study (Kierzkowski and Kisiel 2017a, b) presented the practical application of the simulation model of the security control area at Wrocław Airport. Four different algorithms of dynamic management of resources in the security control area have been proposed in order to ensure the desired system performance. The security control is also considered in terms of reliability (Kierzkowski 2017a). The management aspects of such a process are also addressed (Kierzkowski 2017b). In the paper (Chun and Mak 1999), a stochastic simulation model was developed which allows to adjust the allocation profile of check-in resources for flight schedules of given carriers. In the paper (Bruno and Genovese 2010), the management of the check-in system was proposed to balance the operating costs and waiting time in the queue. In addition to considerations on the possibility of applying static process management with variable efficiency, analyses were also undertaken on the possibility of applying the management of dynamic allocation of check-in desks (Kierzkowski and Kisiel 2007). In the paper (Marelli et al. 1998), a simulation model based on discrete events was developed, which enables the analysis of passenger boarding. The paper (Landeghem and Beuselinck 2002) focuses on the analysis of various passenger boarding strategies. A simulation model was developed for an aircraft equipped with 132 seats in 23 rows. In the paper (Tang et al. 2012) a model was developed in which the dynamics of passenger movement was taken into account. In the article (Kisiel 2020), different passenger boarding strategies and their disruption caused by passengers' behaviour were checked.

The literature review raises important issues:

The analysis of service processes is carried out mainly using simulation models,
The simulation model requires the input of the passenger information stream,
The passport control process has not yet been considered in a dedicated approach.

The aim of this article is therefore to fill the research gap. The article will define the stream of passenger notifications to the passport control process. It will be the basis for future development of the model of this passenger service subsystem. Until now, this subsystem has been taken into account only in a holistic approach. However, due to the limited efficiency of terminals, it must also be properly managed if the air traffic volume after covid-19 epidemic returns to its previous state.

3 Research Background

This chapter will present all the assumptions made during the research. The structure of the analyzed system will also be described. These two factors are extremely important so that those interested can use the results presented in Chap. "In-Depth Evaluation of Reinforcement Learning Based Adaptive Traffic Signal Control Using TSCLAB" for their own system. After all, in an extremely different structure of the system, these data will be inconsistent and will give incorrect indications.

The research was conducted on a system in which the terminal (departure hall) is divided into the implementation of air operations in the Schengen area and the implementation of air operations to countries that do not belong to the Schengen area. All passengers must enter the departure hall through a centralized security control system (Skorupski and Uchroński 2016). After the security control, the passenger is located in the departure hall in the Schengen area. In this zone there are 98% of commercial points (duty-free zone). The process of passport control directs the passenger to the waiting area under the gate for Non-Schengen area. In this zone there is only 1 catering service point. Such a structure does not encourage the passenger to report to the passport control system early. A simplified diagram of the terminal structure is shown in Fig. 1.

The research included recording the time of reporting each passenger to the queue of the passport control system. The place of destination of the passenger was also recorded. This made it possible to estimate the probability density function of passengers' reports by region of destination. The results are presented in Chap. 5.

Destinations were selected on the basis of the destinations that were carried out from Wrocław Airport. These operations were carried out in four main destinations: the United Kingdom of Great Britain, Ireland, Bulgaria and Israel (Table 1).

These destinations can also be translated in other ways. The UK and Ireland are migration destinations, mainly for gainful employment. They are operated by low-cost carriers. Israel can be classified as a mainly holiday destination. However, this destination is still operated by a low-cost carrier as a regular flight. Bulgaria is an example of a typical charter connection contracted by a tour operator.

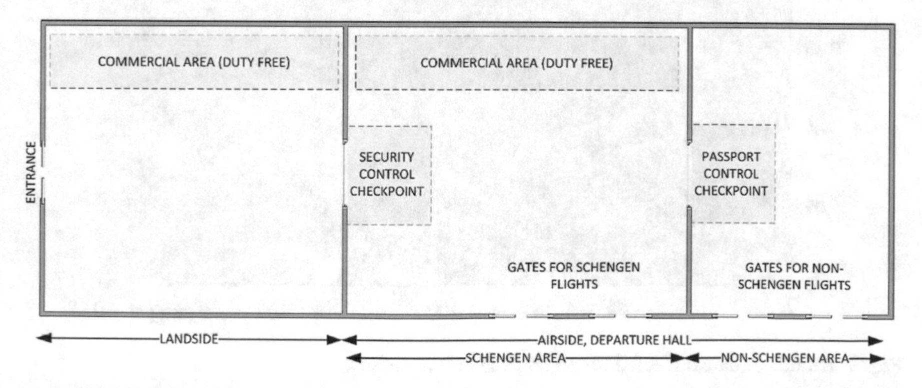

Fig. 1 Terminal structure

Table 1 List of analysed destinations

Region	Code	Destination
United Kingdom Of Great Britain	BFS	Belfast
	EMA	East Midlands
	LUT	Luton
	STD	Stansted
	NCL	New Castle
	MAN	Manchester
	BRS	Bristol
	LPL	Liverpool
Ireland	DUB	Dublin
Israel	TLV	Telaviv
Bulgaria	VAR	Varna

4 Result of Scientific Research

Nearly 20,000 passengers were examined in total. For each group (each region), 30 air operations were tested. The research was conducted in July and August 2019. The results of each of the flights were presented in the form of a probability density functions of passenger reports before the scheduled time of departure of the aircraft. This means that the value 0 on the X-axis indicates the time of departure of the aircraft. Then the marked values on the X-axis indicate the time to scheduled departure. On the Y-axis there is a density of passenger reports to the queuing system.

The results for all operations are shown in Fig. 2. It should be noted that there is a certain area of variation. There are minimum and maximum densities for each X

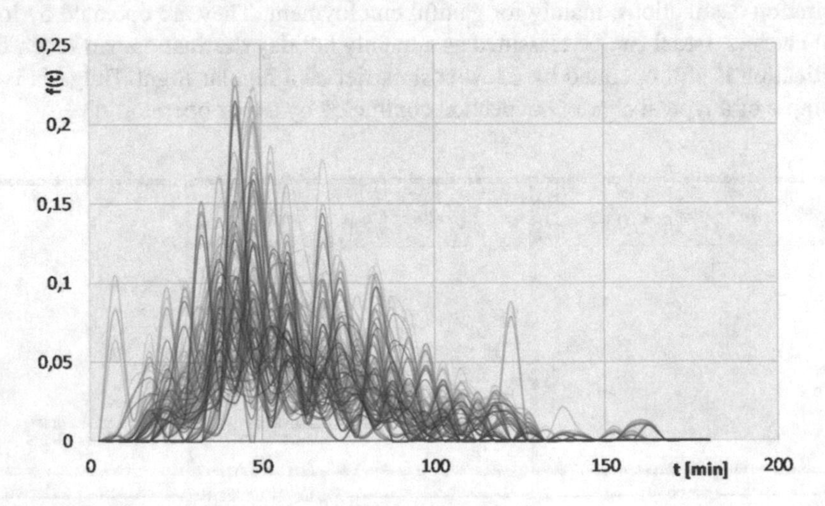

Fig. 2 Results of data collected for all analysed air operations

value. This means that the process is very random. It is therefore advisable to use a simulation model to run a Monte-Carlo simulation when analysing this process.

Therefore, in order for these data to be used in the simulation models, a division into 4 groups (destination regions) was made. Then the mathematical models (theoretical distributions), which can be used in a simulation model, were adjusted.

For each group of connections a theoretical probability density function was estimated. The confidence level equal to 0.05 was assumed. Each of the functions was checked by the Kolmogorov–Smirnov test (Belli et al. 2012). These functions were therefore verified and suitable for use in simulation models. The characteristics of passenger applications flying to the UK with a low-cost carrier for commercial purposes are in accordance with (1)—gamma distribution (shape: 3.8189, scale: 12.642, location: 10.918). For the same assumptions, passengers flying to Ireland report with the function (2)—inverse Gaussian distribution (shape: 320.15, scale: 62.986). On the other hand, passengers in leisure traffic (but still a low-cost carrier) report function (3)—beta distribution (shape1: 2.0232, shape2: 2.8583, boundary: 4.7867, 143.52). Typically charters passengers report with function (4)—inverse Gaussian distribution (shape: 152.21, scale: 48.281).

$$f(t_{UK}) = \frac{(t_{UK} - 10.918)^{2.8189}}{16133.34\Gamma(3.8189)} exp(-(t_{UK} - 10.918)/12.642) \tag{1}$$

$$f(t_{IRL}) = \sqrt{\frac{320.15}{2\pi(t_{IRL})^3} exp\left(-\frac{320.15(t_{IRL} - 62.986)^2}{7934.472(t_{IRL})}\right)} \tag{2}$$

$$f(t_{ISL}) = \frac{1}{\mathrm{B}(4.7867, 143.52)} \frac{(t_{ISL} - 2.0232)^{3.7867}(2.8583 - t_{ISL})^{142.52}}{0.8351^{147.3067}} \tag{3}$$

$$f(t_{BUL}) = \sqrt{\frac{151.21}{2\pi(t_{BUL})^3} exp\left(-\frac{151.21(t_{BUL} - 48.281)^2}{4662.11(t_{BUL})}\right)} \tag{4}$$

The graphical representation of the functions (1–4) is shown in Figs. 3, 4, 5 and 6. Several important conclusions can be drawn on this basis. Firstly, the characteristics for UK and Ireland are almost identical. This suggests that the results are correct. After all, these are 2 destinations that mainly serve the economic migration. Also the same type of carrier (low-cost) operates on these operations. What is very interesting, Fig. 6 shows that passengers who fly for leisure purposes on charter flights, report to the queue much later. This may result from increased interest in the duty-free zone. In turn, Israel, a destination classified as a holiday destination, is a mixture of a low-cost connection with a charter passengers. As a result, the function has been flattened out and, unlike the others, is heading towards a uniform distribution (Fig. 5).

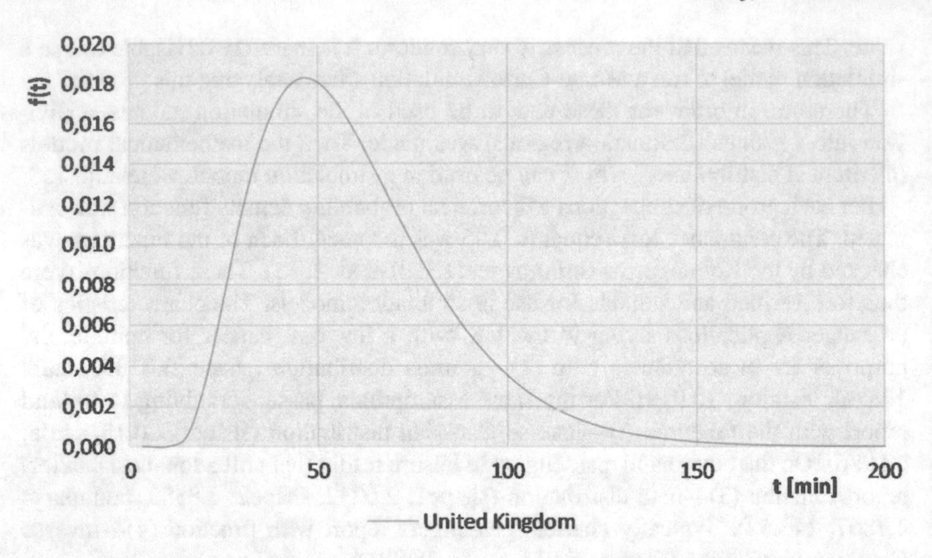

Fig. 3 Theoretical probability density functions of the passenger's reports to the passport control process for the United Kingdom destination

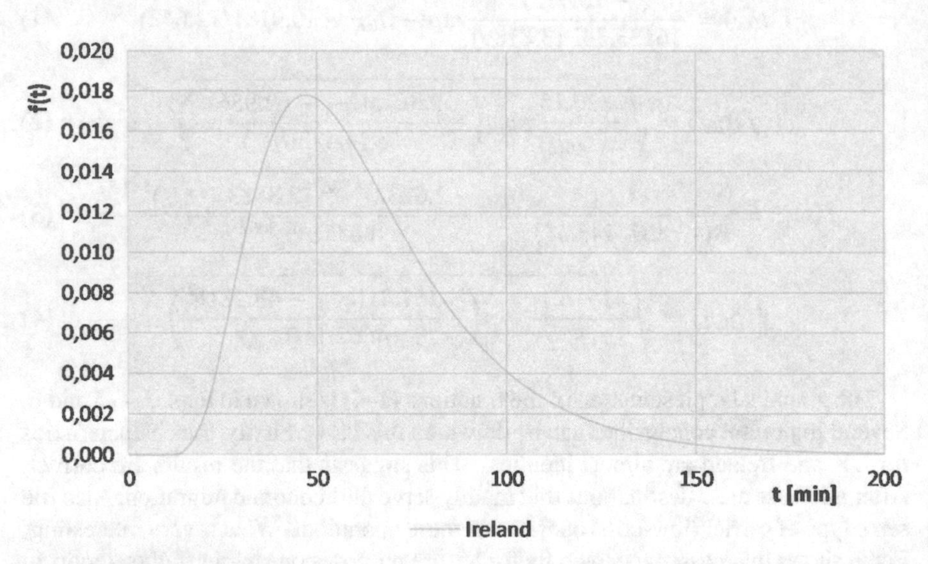

Fig. 4 Theoretical probability density functions of the passenger's reports to the passport control process for the Ireland destination

5 Conclusions

The article presents mainly the results of research, carried out on the real system of passport control at Wrocław Airport. These studies concerned the characteristics of the stream of passenger applications to the queuing system. The aim of the article was

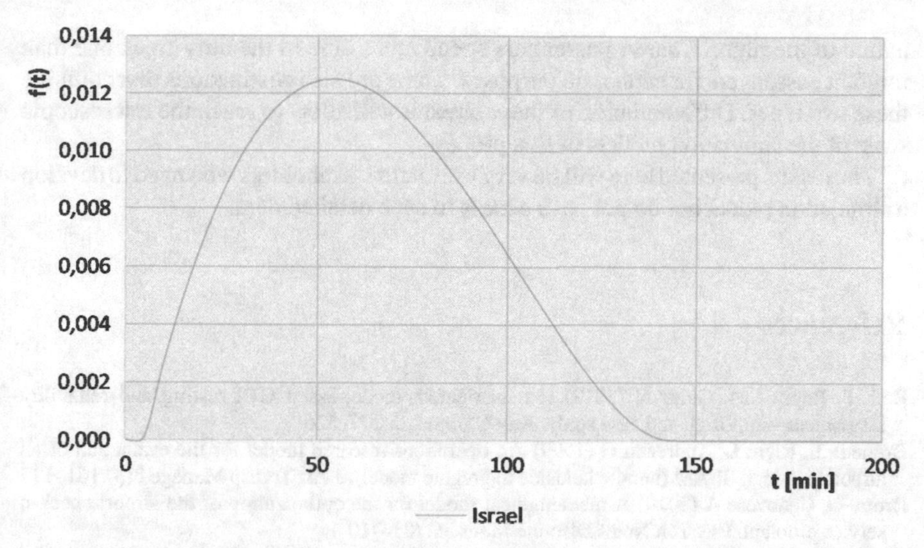

Fig. 5 Theoretical probability density functions of the passenge's reports to the passport control process for the Israel destination

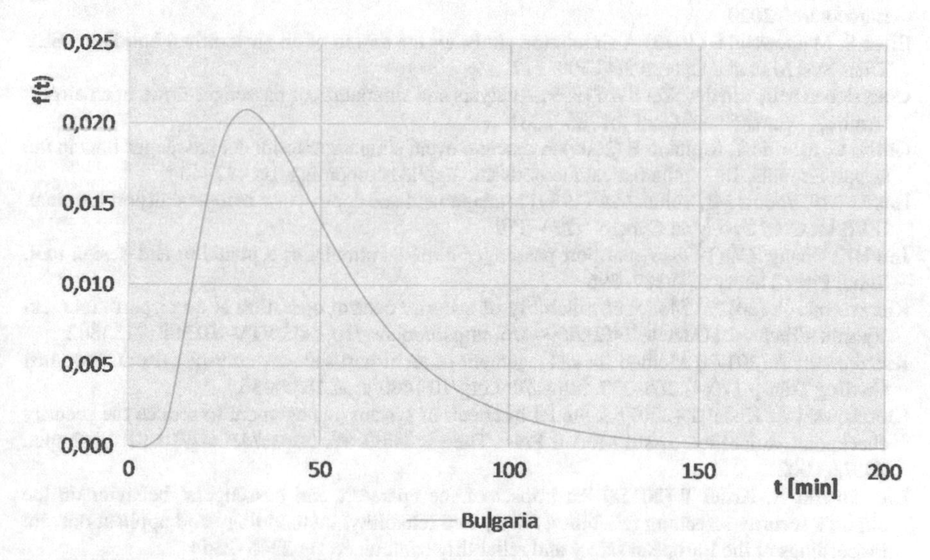

Fig. 6 Theoretical probability density functions of the passenger's reports to the passport control process for the Bulgaria destination

to estimate the probability function, so that it could be used to construct simulation models. The literature review showed that this fragment of passenger service was not yet well explored in a dedicated approach.

The most important finding during the analysis of the results was the conclusion that these distributions are different for the next destinations. This is related to the nature of the flight. Charter passengers spend more time in the duty-free zone than migrant passengers for economic purposes. There are also destinations that combine these two types. Differentiation of these streams will allow to reach the microscopic scale of the simulation models of this process.

The results presented here will be very useful for stakeholders who need to develop a simulation model but do not have access to such detailed data.

References

Belli F, Beyazit M, Güler N (2012) Event-oriented, model-based GUI testing and reliability assessment—approach and case study. Adv Comput 85:277–326

Brunetta L, Righi L, Andreatta G (1999) An operations research model for the evaluation of an airport terminal: SLAM (simple landside aggregate model). J Air Transp Manage 5(3):161–175

Bruno G, Genovese A (2010) A mathematical model for the optimization of the airport check-n service problem. Electron Notes Discrete Math 36:703–710

Chun HW, Mak RWT (1999) Intelligent resource simulation for an airport check-in counter allocation system. IEEE Trans Syst Man Cybern Appl Rev 29(3):325–335

CODA Digest All-causes delay and cancellations to air transport in Europe, Report 2019, Eurocontrol, 2020

Eilon S, Mathewson S (1973) A simulation study for the design of an air terminal building. IEEE Trans Syst Man and Cybern 3(4):308–317

Gatersleben MR, van der Wei SW (1999) Analysis and simulation of passenger flows in an airport terminal. Winter Simul Conf 2:1226–1231

Guizzi G, Murino T, Romano E (2009) A discrete event simulation to model passenger flow in the airport terminal. In: Mathematical methods and applied computing, pp 427–434

Hassan MF, Younis MI, Sultan MA (1989) Management and control of a complex airport terminal. IEEE Int Conf Syst Man Cybern 1:274–279

Jim HK, Chang ZY (1998) An airport passenger terminal simulator: a planning and design tool. Simul Pract Theory 6(4):387–396

Kierzkowski A (2017a) Model of reliability of security control operation at an airport .Tehnički Vjesnik - Technical Gazette 24(2):469–476. https://doi.org/10.17559/TV-20150812153802

Kierzkowski A (2017b) Method for management of an airport security control system. Proc Inst Civ Eng Transp 170(4):205–217. https://doi.org/10.1680/jtran.16.00036

Kierzkowski A, Kisiel T (2007) A model of check-in system management to reduce the security checkpoint variability. Simul Modell Pract Theory 74:80–98. https://doi.org/10.1016/j.simpat.2017.03.002

Kierzkowski A, Kisiel T (2015a) An impact of the operators and passengers' behavior on the airport's security screening reliability, Safety and reliability: methodology and applications. In: Proceedings of the European safety and reliability conference, pp 2345–2354

Kierzkowski A, Kisiel T (2015b) Concept reliability model of the passenger service at the Wroclaw Airport landside area. In: Safety and reliability of complex engineered systems: proceedings of the 25th European safety and reliability conference, ESREL 2015, Zurich, Switzerland, CRC Press/Balkema, pp 1599–1606

Kierzkowski A, Kisiel T (2015c) Functional readiness of the security control system at an airport with single-report streams. In: Proceedings of the tenth international conference on dependability and complex systems DepCoS-RELCOMEX, Brunów, Poland, Springer, pp 211–221. https://doi.org/10.1007/978-3-319-19216-1_20

Kierzkowski A, Kisiel T (2017a) Evaluation of a security control lane with the application of fuzzy logic. Proc Eng 187:656–663. https://doi.org/10.1016/j.proeng.2017.04.427

Kierzkowski A, Kisiel T (2017b) Simulation model of security control system functioning: a case study of the Wroclaw Airport terminal. Air Transp Manage 64, Part B: 173–185. https://doi.org/10.1016/j.jairtraman.2016.09.008

Kisiel T (2020) Resilience of passenger boarding strategies to priority fares offered by airlines. J Air Transp Manage 87:1–9. Doi: j.jairtraman.2020.101853

Kobza JE, Jacobson SH (1996) Addressing the dependency problem in access security system architecture design. Risk Anal 16:801–812

Leone L (2002) Security system throughput modelling. In: 36th annual international Carnahan conference on security technology, pp 144–150

Leone K, Liu R (2003) Measures of effectiveness for passenger baggage security screening. Transport Res Record J Transport Res Board 1822:40–48

Li Y, Gao X, Xu Z, Zhou X (2018) Network-based queuing model for simulating passenger throughput a tan airport security checkpoint. J Air Transp Manage 66:13–24

Ma W, Fookes C, Kleinschmidt T, Yarlagadda P (2012) Modelling passengers flow at airport terminals. Individual agent decision model for stochastic passenger behaviour. In: Proceedings of the 2nd international conference on simulation and modeling methodologies, technologies and applications (SIMULTECH-2012), pp 109–113

Marelli S, Mattocks G, Merry R (1998) The role of computer simulation in reducing airplane turn time, technology/product development. AERO Mag 01, Boeing

Neufville R (2008) Building the next generation of airports systems. transportation infrastructure 38(2)

Newell GF (1971) Application of queuing theory. Chapman and Hall, London

Schultz M, Fricke H (2011) Managing passenger handling at airport terminals. In: Ninth USA/Europe air traffic management research and development seminar. Berlin, Germany, pp 1–10

Skorupski J, Uchroński P (2016) A fuzzy system to support the configuration of baggage screening devices at an airport. Expert Syst Appl 44:114–125

Solak S, Clarke J-PB, Johnson EL (2009) Airport terminal capacity planning. Transport Res Part B 43:659–676

Tang T-Q, Wu Y-H, Huang H-J, Caccetta L (2012) An aircraft boarding model accounting or passengers' individual properties. Transport Res Part C: Emerg Technol 22:1–16

van Boekhold J, Faghri A, Li M (2014) Evaluating security screening checkpoints for domestic flights using a general microscopic simulation model. J Transport Secur 7(1):45–67

van Landeghem H, Beuselinck A (2002) Reducing passenger boarding time in air-planes: a simulation based approach. Eur J Oper Res 142:294–308

Situation in Railway Sidings Operation in Slovakia Based on the Selected Criteria

Lenka Černá, Vladislav Zitrický, and Jozef Gašparík

Abstract The railway transport sector in Slovakia has recorded significant changes in recent decades, which were related to the change in the orientation of the national economy as well as the creation of a single European railway market. The transformation of the national economy and the railway market has also had an impact on the operation of railway sidings. The number of active railway sidings gradually decreased and basically copied the development trends in the area of the share of railway transport in the transport market. However, the position of railway sidings in the railway transport system still has a significant share today and it is therefore necessary to know the effects of various factors in railway transport on their operation. Many railway sidings have a valid operating licence however, their transport performance is very poor or none. On the other side there are railway sidings with expired operating licence which abolition is protracted, financially and legally demanding. Nevertheless, the railway siding is very important part of rail freight transport operation. The paper focuses on the examining of the railway siding performance share in overall rail transport performance and it contains dependence analysis of railway siding performance to selected criteria: total transport volume in international transport, transport price and number of railway siding service. Partial goal of paper is the analysis of active and no-active railway sidings.

Keywords International transport performance · Railway transport · Railway siding · Dependence analysis

L. Černá · V. Zitrický (✉) · J. Gašparík
Department of Railway Transport, Faculty of Operation and Economics of Transport and Communications, University of Žilina, Žilina 010 26, Slovak Republic
e-mail: vladislav.zitricky@fpedas.uniza.sk

L. Černá
e-mail: lenka.cerna@fpedas.uniza.sk

J. Gašparík
e-mail: jozef.gasparik@fpedas.uniza.sk

M. Petrović and L. Novačko (eds.), *Transformation of Transportation*, EcoProduction,
https://doi.org/10.1007/978-3-030-66464-0_7

1 Introduction

Increasing international integration is one of the characteristics of the world economy and this integration process results in the globalization of international trade. Transport is a significant part of international trade and railway transport in particular has an irreplaceable position in the world economy (Veličković et al. 2018). Railway sidings are an important part of rail transport and they have a significant share of the rail transport performance. Railway sidings in the Slovak Republic account for about 70% of the total rail transport volume (Abramovič et al. 2014).

Currently the importance and position of railway sidings not to lose its importance in railway transport. The performance of rail transport is largely made up of transports performed on railway sidings and therefore the area of siding operation must be constantly developed and adapted to the conditions of the competitive environment of the transport market (Zitrický 2016).

The declining trend of railway transport using in the EU was also reflected in the territory of the Slovak Republic and it also affected the performance in railway siding operation. The number of operated railway sidings is continuously decreasing over the years after the change of the orientation of the national economy to the principles of the market economy. The unfavorable development of the use of railway transport has not been stopped even by the adoption of regulatory measures by the European Commission, and therefore it is necessary to identify new possibilities for the development of railway transport and look for possibilities for its development (Hricišonová 2017).

The Slovak economy is part of the world economy and the behaviour of world markets influence whole areas of the national economy, including railway transport. The economy of the Slovak Republic saw a transformation process from a planned economy to a market economy (Dolinayová et al. 2016). This caused changes in the transport market and resulted in railway transport losing its position in the market and national economy. The transformation process of the economy decreased the number of railway sidings, but they didn't lose their position in the railway market (Ližbetin and Stopka 2020). The importance of railway sidings as a part of railway transport is shown in the correlation analysis of the rail transport performance, costs in rail transport and performance. Research is oriented on the Košice region in Slovakia.

2 Globalization and the Transport Market

The International Monetary Fund defines globalization as a historical process, the result of human innovation and technological progress. Globalization means the growing integration of global economic dependence. The dependence of the global economy is carried out through the transaction of goods, services and economic capital crossing international borders. Križanová et al. (2014); Buková et al. (2016) said that up until 2019, the leaders of global economic growth will be countries and markets from Asia and Africa. The Slovak economy will still depend on European Union trade, as was the case in 2016 when exports from Slovakia to the EU made up 82.81% of the total while imports accounted for 62.5% (Buková et al. 2016).

Transport, which is one of the sectors in the EU, creates gross added value to an economy. In his study, Horniakova says that transport in 2011 had a 4.7% share of the total added value of the EU expressed at the current prices. Based on the same study, the most important impact on the added value of the transport mode is provided by land transport (road transport, rail transport and pipeline transport). Within the land transport market, road transport (about 70–80%, it depends on the EU county) accounts for the biggest share, followed by rail transport and pipeline transport. (Francesco et al. 2016) The position of railway transport in the market in the Slovak Republic is the almost the same as in other EU member states. The performance based on transport volume is about 20% in the last ten years. (Horniaková 2014).

Transport plays an important role in the process of globalization. Its primary task is to ensure the transfer of goods from the place of production to the place of consumption, due to the difference in the occurrence of production raw materials, their processing and final consumption. A significant part of the input raw materials intended for the production and transformation process is located in various places. The task of transport is the fastest and most efficient transport of goods, raw materials by means of transport along transport routes. (Černá et al. 2017).

The relationship between transport and globalization can be expressed in terms of the share of freight transport in relation to gross domestic product (GDP). Using this indicator, we can observe how the volume of freight transport developed compared to GDP. (Černá et al. 2017).

In the Fig. 1 the correlation relationship is shown and the correlation coefficient between the gross domestic product of the Slovak Republic and total transport volume in Slovakia is calculated.

Based on the calculated correlation between the gross domestic product and total transport volume in the Slovak Republic we can claim a very strong dependency. Correlation coefficient has a value 0.999 and the Pearson's coefficient is 1.00. Variability of the gross domestic product is 99%. The result of correlation analysis shows that the gross domestic product and total transport volume are changing their direction. In this case it is obvious: any increase of the gross domestic product means the growth of transport performance (Černá et al. 2017).

3 Development Analysis of Railway Sidings in the Slovak Republic

In the past, railway transport was the most used transport mode in terms of passengers and freight. Railway transport started to lose its position in the transport market in the global economy in the second half of last century (Kudláč et al. 2017). Road transport has taken over its position in the market. The Slovak economy and market copied the

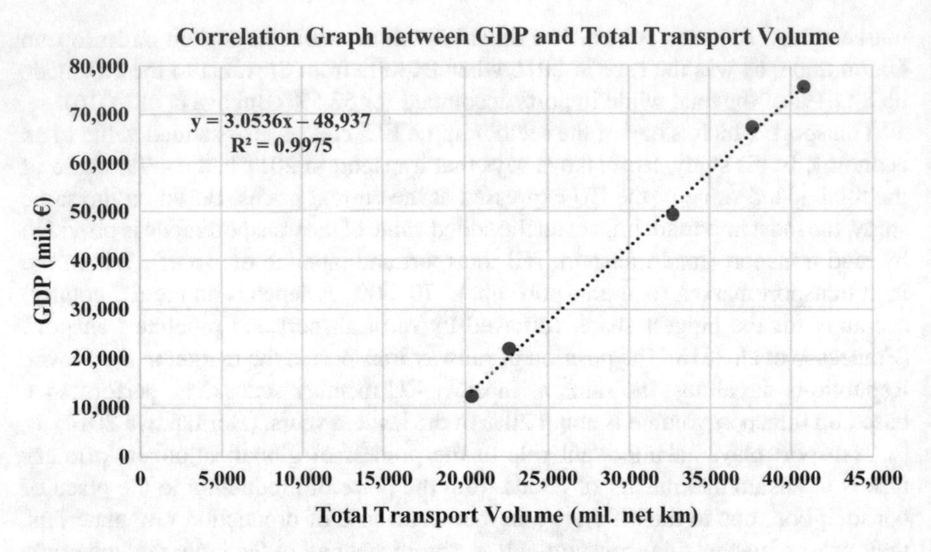

Fig. 1 Relation between total transport volume and the gross domestic product. *Source* (Černá et al. 2017)

trend in the world economy after the transformation process in 1989, when the Slovak economy became part of the global market economy. This process saw less railway sidings in operation. Railway sidings in the past were built near great manufacturing facilities and the majority of production was transported by rail transport. The current trend of using railway sidings is oriented on their destruction. Railway sidings are no longer in the rail process because their operation is mostly inefficient and costly. Another problem is the high cost to build new railway infrastructure near the new logistic centres (Zitrický et al. 2016).

Figure 2 shows the number of infrastructure operational licences (railway siding operation licences) in the Slovak Republic between 1997 and 2015. These licenses were issued by the Transport Authority of the Slovak Republic under the railway infrastructure law.

After 1998 the number of issued licences for the operation of railway sidings decreased due to falling demand. The first small change can be seen in 2002 and this was the year that the Railways of the Slovak Republic were transformed by the legislative of the EU. The second small change in the negative trend was cause by the full opening of the rail market to competition. Figure 3 shows the validity license of railway siding operations in 2016 in the Slovak Republic.

In the Slovak Republic there are 772 railway sidings which don't have a valid license for operation, which in turn means they are not being operated. Some 222 railway sidings have a valid license for operation while eight railway sidings are in bankruptcy. The low number of operated railway sidings is due to the influence of the global market and the structure of the Slovak national economy that are characterized above (Hricišonová 2017).

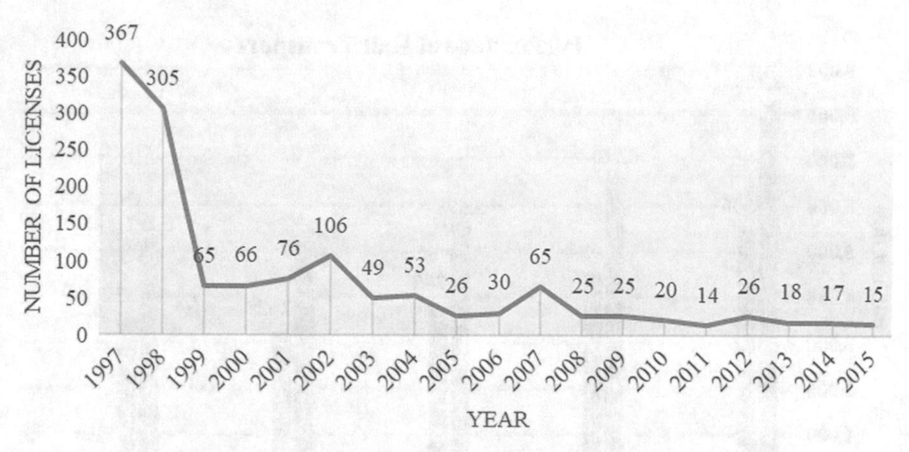

Fig. 2 Development of the number of licenses issued for the operation of the railway in the SR. *Source* Transport authority of SR

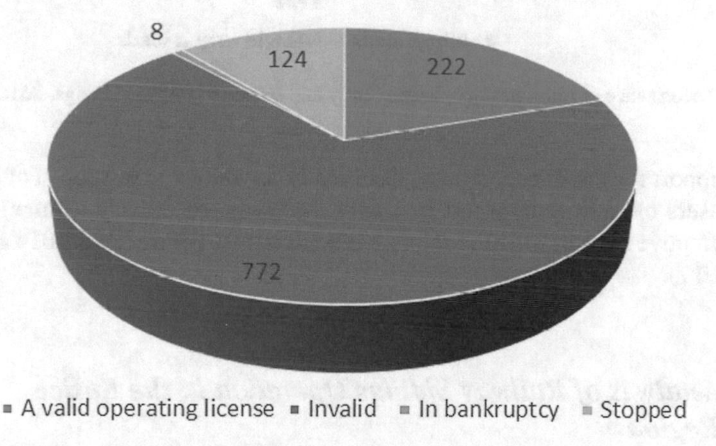

Fig. 3 Validity licenses of railway siding operations. *Source* Hricišonová (2017)

The number of railway sidings has had a decreasing slope. Despite of this fact railway sidings still have a majority share on the rail transport performance. In the Fig. 4 we can see a share between single wagon loads and block trains which are operated between railway sidings (ŽSR 2014 and Černá et al. 2017).

Based on the Study and Survey of the Railway of Slovak Republic on railway sidings that was realised between the owner of railway sidings and carriers we can see an interesting usage of railway sidings. However, the survey also showed a great weakness of this transport type. It revealed the weakness on the side of an infrastructure manager, railway undertakings (national and private) and also of the owners of railway sidings. The most relevant weaknesses are: data update by means of communication, efficient usage of information and update of transport condition (technical and operation). Another great problem was identified in the form of a small

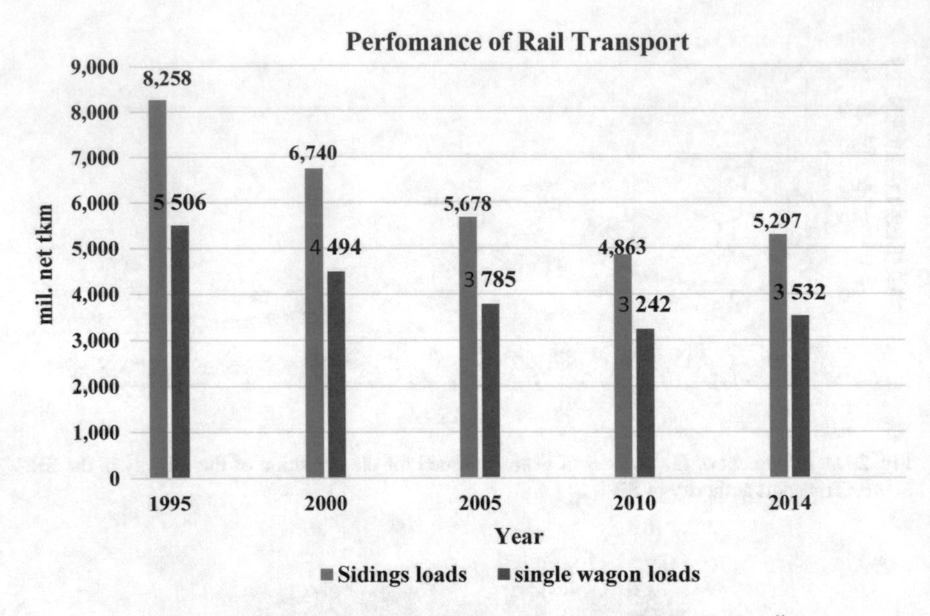

Fig. 4 Performance of rail transport. *Source* Study and survey of railway sidings, ŽSR

state support for the transport area, specifically for railway transport. For example, many users of railway transport (not only the owner of railway sidings) expect a more effective support from the state for the rail infrastructure (ŽSR 2014 and Černá et al. 2017).

3.1 Analysis of Railway Sidings Operation in the Košice Region

Organizational unit The Regional Directorate Košice is a separately accounting organizational unit belonging to the Railways of the Slovak Republic. The Regional Directive Košice is responsible for the entrusted assets, receivables and liabilities of the Railways of the Slovak Republic within its district and follows the financial plan. The costs are covered by internal revenues, external revenues and a contribution from the General Directorate of the Railways of the Slovak Republic (Csatlosova 2012).

The subject of activities of the Regional Directorate Košice is to a limited extent. Their main tasks include:

Management, organization and coordination of safe and smooth transport activities,
Providing organizational prerequisites for fulfilling orders of rail users for the passenger and freight transport,
Ensuring the operability of the railway infrastructure,

Management and organization of the administration of railway lines and constructions, administration of entrusted equipment
Management, organization and provision of operations,
Ensuring an optimal and continuous supply of gas and heat for all organizational units of ŽSR from foreign suppliers,
Operation of fixed electric traction and high-current equipment, other specialized electrical equipment. (Csatlosova 2012).

In the Košice region, 93 railway sidings were registered in 2016. These registered railway sidings are directly connected to the infrastructure of the national infrastructure manager—ĪŽSR (Railways of the Slovak Republic). Railways sidings that are directly connected to the national rail infrastructure are called "railway sidings of first sequence". The national infrastructure manager in the Slovak Republic only registers the first sequence of railway sidings. Railway sidings that are connected to the infrastructure of another railway siding (railway siding of second or third sequence) are not registered in the database of ŽSR. Figure 5 shows the percentage share of active railway sidings in the Košice region. There are 16 non-active railway sidings from the first sequence, while another 77 railway sidings are active (Hricišonová 2017).

Figure 6 shows the geographic point of view of railway sidings in Košice region. On the map are shown this private sidings alongside to the railway lines.

The most transported commodities on the selected railway sidings in Košice region are wood and minerals (magnesite, limestone, and dolomite). After that, the next most transported commodities are gas, metallurgical products, intermodal transport units, foodstuff, chemical products and other nonspecific goods. Companies that used railway sidings only for the unloading of goods, move their produced products via road transport. Figure 7 shows the percentage share of commodities that are transported on railway sidings in the Košice region. (Hricišonová 2017). Figure 8 shows the percentage distribution of railway sidings which are oriented towards loading or unloading operations, including ones which carry out both operations. 37 railway sidings have common loading operations, while there are 32 railway sidings that just have loading operations and 8 railway sidings just for unloading operations.

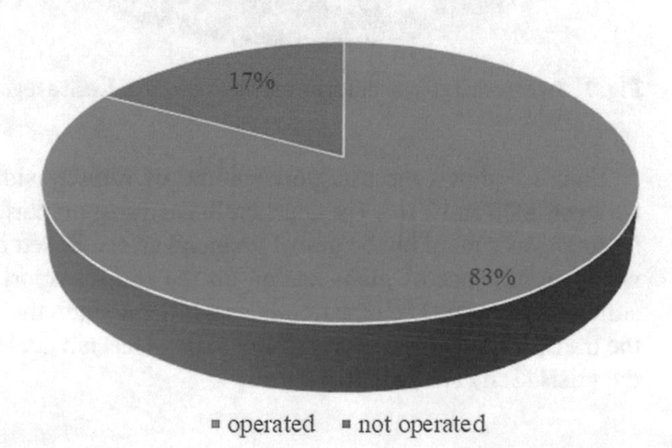

Fig. 5 Percentage share of active railway sidings in the Košice region. *Source* Hricišonová (2017)

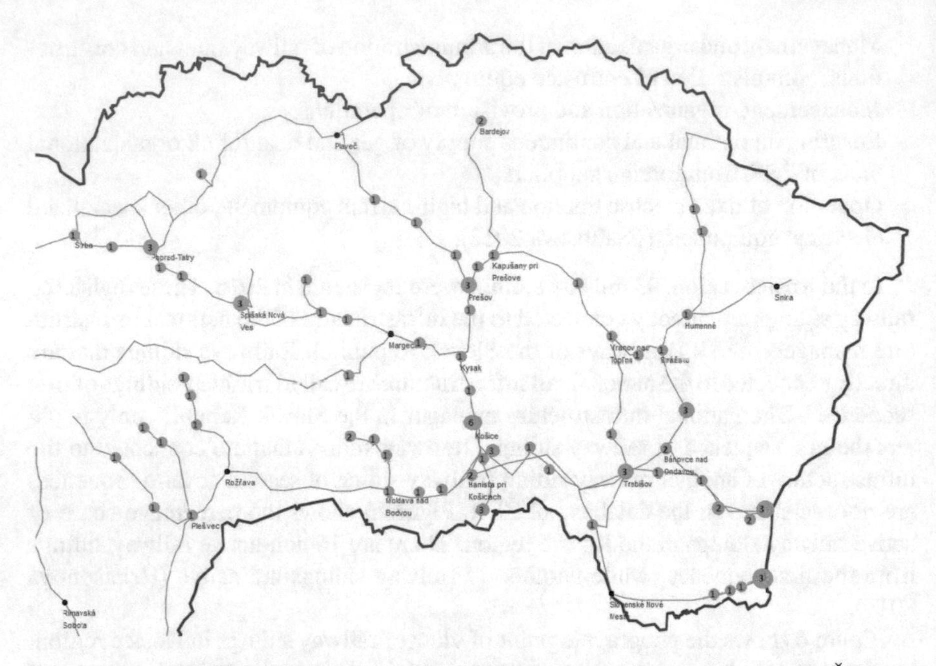

Fig. 6 Geographical location on railway sidings in Košice region. *Source* authors on the ŽSR map

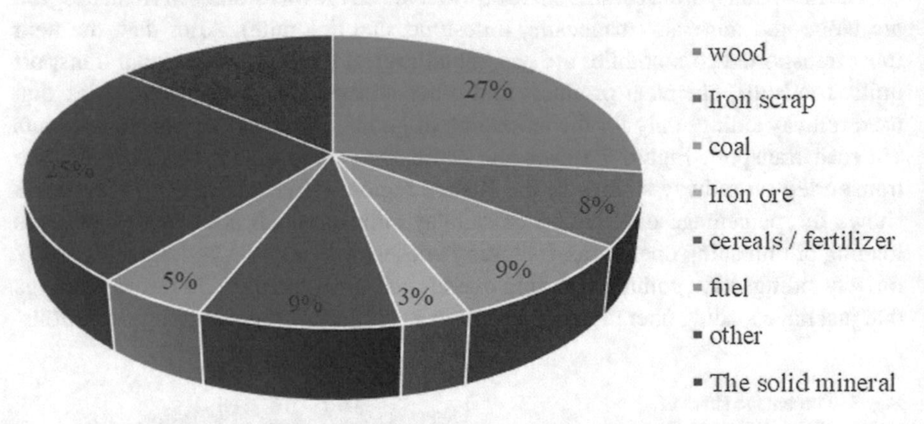

Fig. 7 Transported commodities on railway sidings in Košice region. *Source* Hricišonová (2017)

Figure 9 shows the transport volume of railway sidings in the Košice region between 2008 and 2016. The decrease in the transport performance of railway sidings in 2009 was caused by the global financial crisis. Based on this fact, we can see the negative influence of globalization on the rail transport performance and railway sidings operation. The situation has stabilized since the global financial crisis, but the transport performance on railway sidings has not achieved the values seen before the crisis (Schwartzová 2013).

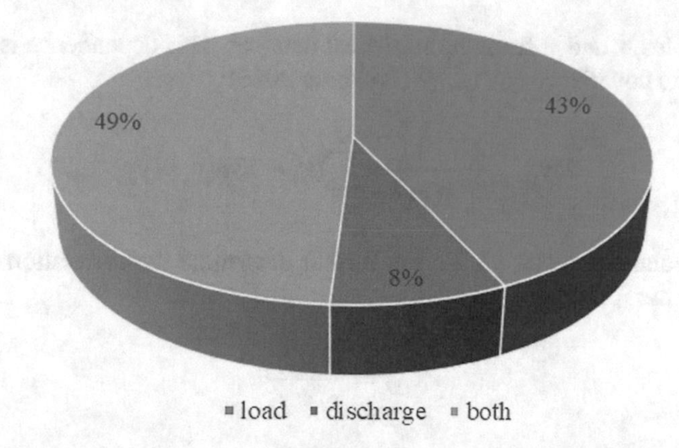

Fig. 8 Percentage view of the method of handling on railway siding in Košice region. *Source* Hricišonová (2017)

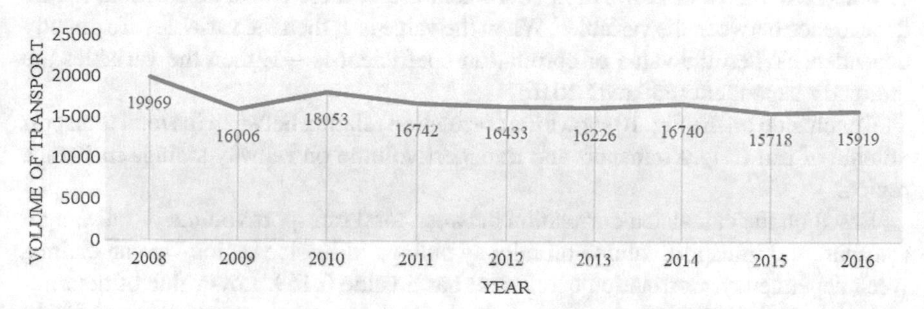

Fig. 9 Volume of transport on railways sidings in Košice region (thousands tonnes). *Source* Hricišonová (2017)

3.2 *Dependence of Railway Sidings Performance on the Basis of Selected Indicators in the Kosice Region*

Dependence analysis is oriented on the relation between rail transport price and transport volume on the railway sidings, number of railway sidings in the Košice region and the total transport volume of rail freight transport and between total transport volume of rail freight transport and transport volume on railway sidings in the Košice region.

Correlation analysis is used for the dependence analysis of the railways sidings performance and selected criteria (Palkovič 2016).

Based on Benko's definition "Correlation is a mutual linear relation (dependence) of two random variables X and Y. This relation can be direct (if one variable increases, then the second variable increases, too), or indirect (with one variable growth there is a fall of the second one)". The mathematical calculation, which results in numerical data showing the dependence of two or more elements of a statistical data-set, is called a correlation number. The correlation coefficient r(x, y) represents the dependency of two variables, x and y, from the statistical data-set. This dependence is expressed with a statistic covariance cov (x, y) (Palkovič 2016):

$$\mathrm{cov}_{(x,\,y)} = \frac{1}{n-1} * \sum_{i=1}^{n} (x_i - \overline{x}) * (y_i - \overline{y}) \tag{1}$$

After calculating cov (x, y), we can further determine the correlation coefficient according to the formula:

$$r_{(x,y)} = \frac{k_{(x,y)}}{s_x * s_y} \tag{2}$$

Palkovič says the value of correlation coefficient expresses a linear degree of dependency of variables x and y. The value of correlation coefficient is from −1 to 1. When the values of correlation coefficient are 0, there exists no relation, i.e. no dependence between the variables. When the value is 1, then the variables are directly dependent. When the value of correlation coefficient is −1, then the variables are indirectly dependent (Palkovič 2016).

Conclusion on the Fig. 10 shows the correlation relation between the total transport volume of rail freight transport and transport volume on railway sidings in Košice region.

Based on the calculated correlation between total transport volume of rail freight transport and transport volume on railway sidings in Košice region we can claim a weak dependency. Correlation coefficient has a value 0.169. Low value of determination coefficient (2.86%) shows weak dependency between performance on railway

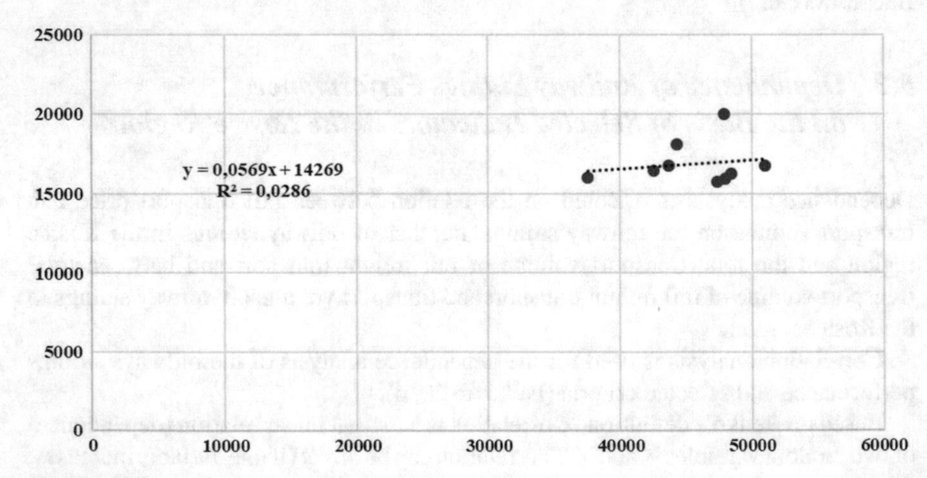

Fig. 10 Relation between total transport volume and the transport volume on railway sidings. *Source* Authors

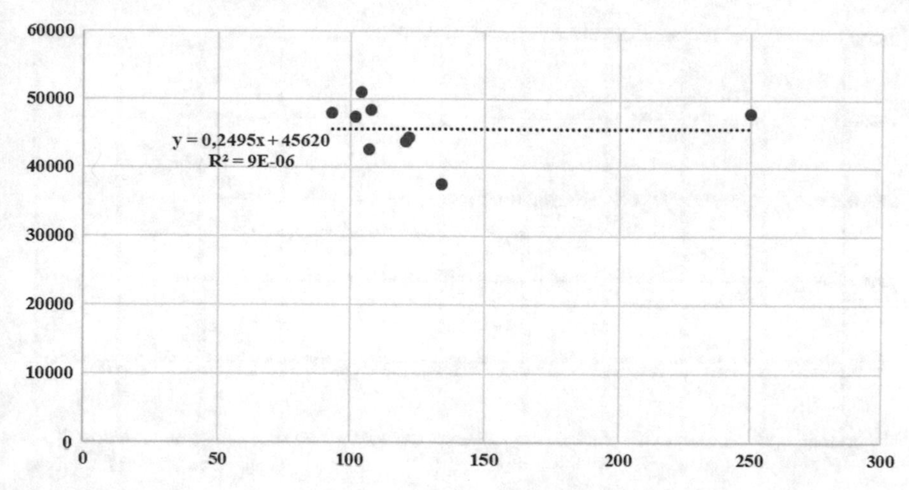

Fig. 11 Relation between number of siding and the transport volume on railway sidings. *Source* Authors

sidings and total transport volume in rail transport. This fact can be caused by comparison only by performance in Košice region to total transport performance in Slovak republic.

In the Fig. 11 is a correlation analysis between the number of railway sidings in Košice region and total transport volume of rail freight transport.

Based on the calculated correlation between the number of railway sidings in Košice region and total transport volume of rail freight transport we can claim a medium dependency. Correlation coefficient has a value 0.003. Almost null value of determination coefficient means no dependency between transport performance on railway sidings and number of railway sidings. Results misrepresent the railway siding position in rail transport mode because based on study (Zitrický et al. 2016) percentage share of railway sidings in the rail transport volume is about 40%.

In the Fig. 12 is a correlation analysis between the transport volume of railway sidings in Košice region and rail transport price. Based on the correlation coefficient (−0.662) we can say that the dependency between transport price and transport volume of railway sidings is great. Based on the value of correlation coefficient we see indirect dependency. In this case the results of correlation analysis predict that low price level of rail transport influences transport volume on the railway sidings. Variability of transport performance on railway sidings is 43.9%. This value shows middle dependence between compared variables.

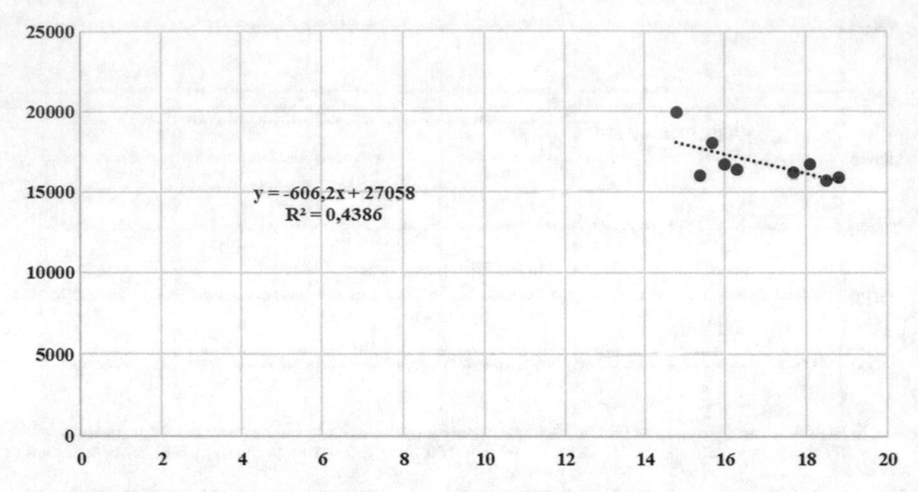

Fig. 12 Rail transport price and transport volume on the railway sidings. *Source* Authors

4 Proposal for Measures to Increase the Efficiency of the Railway Sidings Operations

The proposal of measures to increase the efficiency of the use of railway siding operation is necessary to focus on the possibilities of the development of railway siding operation itself and the increase of transport performance on the railway sidings. The declining trend in the use of railway sidings must be stopped, especially from the point of view of sustainable development and the reduction of road infrastructure congestion.

4.1 *State Support of Railway Sidings Operation*

The international position of the Slovak Republic calls for pressure on individual line ministries to take regulatory measures to ensure sustainable development and rational use of natural resources, especially with a focus on environmental issues.

Transport, as one of the biggest air pollutants, faces constant pressure from environmental environments and, on the other hand, seeks to secure demand for transport services, which has been on an ever-increasing trend over the decades. The principles of the free market do not allow states to intervene directly in a competitive environment or in the transport market. Therefore, the position of the state is narrowed to create a harmonized environment for individual modes of transport. All interventions in the market environment must be carried out sensitively and must have at least a societal effect.

The support of railway operation, i.e. also of railway sidings, can be understood as society-wide acceptable, at least in the form of a decrease in congestion in road transport and improvement of environmental protection. With appropriate support from the state, it is possible to reverse the unfavourable state of siding operation in the last 20 years in the Slovak Republic.

A specific option for developing siding operation is support for the reconstruction and renewal of non-operating sidings from EU funds, the financial coverage of EU funds, which means that all funds spent must serve the public.

The proposal for the reconstruction and renewal of sidings from the EU funds must therefore necessarily mean that sidings financed through the EU must be accessible to the public on a non-discriminatory basis. In essence, this case would involve the construction of new freight stations, which would serve carriers that do not have sufficient funds and do not like the loading of goods on the load track in the station according to the current conditions of the carriers. Railway sidings supported by EU funds could form small terminals that are able to meet the transport needs of the micro-region in which they are located. This proposal is applicable not only in the district of OR Košice but basically on the entire network of Slovak railways.

4.2 Design of Railway Sidings Operation Information System for an Infrastructure Manager

The processing of databases, or their non-existence, from the documents of the infrastructure manager proved to be a problematic place in the analysis of the obtained data of the siding operation in the district of Košice.

The Railways of Slovak republic maintains a database of valid siding contracts as well as in the operating rules of railway stations they have the characteristics of all sidings that occur in a given station, but it is problematic to determine whether a given siding performs transport services or not.

In the case of railway sidings of the first sequence (railway sidings connected to the national railway network), it is not problematic to determine their transport performance, or the operation of the railway siding capacity due to the existence of the Contract on Railway Interconnection. The critical place at the railway sidings is the so-called second and third sequences, which these contracts do not have and, in essence, the infrastructure manager is not able to determine exactly whether the railway siding is in operation or not.

An information system designed for railway siding operations will significantly simplify access to the necessary information about a specific siding needed for the infrastructure manager. The minimum requirements from the user's point of view that the information system must contain are:

- number of operated railway sidings,
- number of non-operated railway sidings,
- validity of the railway interconnection contract,

- names of the railway siding owner,
- railway siding operator,
- the carrier that operate on the railway sidings.

5 Conclusion

Globalization is a process that influences many economics sectors. The globalization trend supported the growth of the Slovak economy after the transformation process in 1989. However, the impact of globalization hasn't been the only positive effect. For rail freight transport it has meant a decrease in transport performance and in many cases the termination of railway. Nevertheless, railway sidings have their importance and a significant position in rail transport. Nevertheless, the correlation in two cases doesn't show dependency, but we can not to say that the railway sidings haven't the important position in railway operation. Globalization influence is representing in relation between transport price and transport performance on railway sidings and there are several possibilities to support this. The first way is state subvention of the railway infrastructure, especially railway sidings. The second method is legislative changes concerning the building of logistic centres. Part of the Slovak economic transformation was the building of many logistic centres, but only very few of them are connected to the railway infrastructure (Fedorko et al. 2018).

Acknowledgements The paper was supported by the VEGA Agency, Grant No. 1/0509/19 "Optimizing the use of railway infrastructure with support of modal split forecasting", at Faculty of Operations and Economics of Transport and Communication, University of Žilina, Slovakia.

References

Abramović B, Brnjac N, Škrinjar JP (November 2014). Railway industrial track as the last mile in supply chain management. In: Proceedings of the second international conference on traffic and transport engineering (ICTTE), Belgrade. 27–28

Buková B, Brumerčíková E, Kondek P (2016) Determinants of the EU transport market. In: ESM 2016: the 2016 international conference on engineering science and management, Zhengzhou, Henan, China. Paris: Atlantis Press

Černá L, Zitrický V, Blaho P (2017) The impact of globalization on the performance of the railway sidings operation based on selected indicators. In: 17th International scientific conference: globalization and its socio-economic consequences, proceedings part I. Žilina

Csatlosova E (2012) Analysis of the operation of railway sidings in the district of the regional directorate of ŽSR Košice. Bachelor thesis, University of Žilina

Dolinayová A, Čamaj J, Daniš J (2016) Evaluation of investment efficiency in the new database solution for rail freight transport in the context of globalization. In: Globalization and its socio-economic consequences-16th international scientific conference, University of Žilina, Žilina

Fedorko G, Molnar V, Honus S, Neradilova H, Kampf R (2018) The application of simulation model of a milk run to identify the occurrence of failures. Int J Simul Model 17(3):444–457. https://doi.org/10.2507/Ijsimm17(3)440

Francesco R, Gabriele M, Stefano R (2016) Complex railway systems: capacity and utilisation of interconnected networks. Eur Transp Res Rev 8:29. https://doi.org/10.1007/s12544-016-0216-6
Horniaková T (2014) Analysis of the impact of globalization on the performance of the railway company Cargo Slovakia, a. s. in international rail transport, Bachelor thesis, University of Žilina
Hricišonová P (2017) Analysis of siding in Košice region, Bachelor thesis, University of Žilina
Križanová A, Gajanová Ľ, Masárová G (2014) The need of competitive intelligence implementation in transport enterprises in the Slovak republic. In: Transport means 2014: proceedings of the 18th international conference. Kaunas University of Technology, Lithuania, Kaunas
Kudláč Š, Štefancová V, Majerčák J (2017) Using the saaty method and the FMEA method for evaluation of contraints in logistics chain. Procedia Eng 187:749–755
Ližbetin J, Stopka O (2020) Application of specific mathematical methods in the context of revitalization of defunct intermodal transport terminal: a case study. Sustainability 12:2295. https://doi.org/10.3390/su12062295
Palkovič P (2016) Regress and correlation analysis. Available from Internet: https://spu.fem.uniag.sk/cvicenia/ksov/palkovic/statistikaB/prednasky/8.Regresná%20a%20korelačná%20analýza.pdf
Schwartzová D (2013) The Košice regional directorate is also the largest number of people, the railway traffic light. XXIII, 4
Veličković M, Stojanović Đ, Nikoličić S, Maslarić M (2018) Different urban consolidation centre scenarios: impact on external costs of last-mile deliveries. Transport 33(4):948–958. https://doi.org/10.3846/16484142.2017.1350995
Zitrický V, Černá L, Ponický J (2016) Possibilities of development of railway sidings in the Slovak Republic, ICTTE 2016 international conference on Traffic and transport engineering. Belgrade, Serbia, Belgrade
ŽSR (2014) Case study and survey about the revitalisation of railway sidings on the Slovak railways. Železnice Slovenskej republiky, Bratislava, p 2014

Applying Multi Criteria Analysis in Evaluation of Distribution Channels

Karla Brezović, Ratko Stanković, Mario Šafran, and Goran Kolarić

Abstract Optimal allocation of available distribution channels has major importance in distributing products to the market. Allocation is made against the relevant set of criteria that have different importance and according to the requirements of the products. For these reasons, a proper evaluation of the available distribution channels is crucial for effectiveness and efficiency of distribution of products. This paper outlines the possibilities of applying AHP Method of multi criteria analysis in evaluation of optional distribution channels, in distribution of confectionery products.

Keywords Distribution channels · Multi criteria analysis · AHP method · Evaluation

1 Introduction

Distribution channel decisions are the most significant decisions of the company in terms of making products and services available for the consumer or business user who needs it. Well-chosen distribution channels affect all other marketing decisions: decisions on the sale of certain goods, place of sale, participants in the sale, pricing policy and economic decisions (Matić 2016).

Competitive advantages are created through distribution channels, such as lower logistic costs compared to the competitors, better market coverage, proximity of products to customers, better service, faster delivery.

K. Brezović · R. Stanković (✉) · M. Šafran · G. Kolarić
Faculty of Transport and Traffic Sciences, University of Zagreb, Vukelićeva 4, 10000 Zagreb, Croatia
e-mail: ratko.stankovic@fpz.unizg.hr

K. Brezović
e-mail: karla.dubr@gmail.com

M. Šafran
e-mail: m.safran@fpz.unizg.hr

M. Petrović and L. Novačko (eds.), *Transformation of Transportation*, EcoProduction,
https://doi.org/10.1007/978-3-030-66464-0_8

The main idea of the paper refers to evaluation of the existing distribution channels for the manufacturer of confectionery products in Croatia, suggestion for an additional distribution channel, and final ranking of all distribution channels by using multi criteria analysis.

The importance of distribution channels is analyzed both for individual economic operators, groups and for the aggregate national economy. This is why earlier concepts of the role and importance of commerce in the national economy are losing their importance (Segetlija et al. 2011).

In recent years, the chocolate and confectionery industry has become particularly important in the food industry. It is characterized by high average profitability and constant annual growth rate. The attractiveness of the industry has increased in the intensity of competition and the offensive actions of chocolate companies. Adopting the appropriate organization strategy at the headquarters and domain level is a necessary condition for efficient and effective operation in a volatile economic environment. Confectionery companies should not only take care of the attractiveness of the manufactured products but also attempt to maintain a competitive international position (Mierzwa and Zimmer 2017).

The AHP method was applied in evaluation of distribution channels in the case study of the confectionary manufacturing company, outlined in this paper. The major characteristic of the AHP method (developed by prof. Thomas Saaty, University of Pittsburgh, in 1970s) is the use of pairwise comparisons, which are used both to compare the alternatives with respect to the various criteria and to estimate criteria weights. The AHP method was quite common in the literature reviewed during this study.

E-commerce was added as a suggestion for an additional distribution channel. The Internet provides retailers with potentially powerful opportunities to boost sales, increase market share, and generate new business through new services. One of the challenging questions that retailers are facing in that respect is how to organize the physical distribution processes during and after the transaction has taken place (Koster 2003).

The most favourable distribution channel are ranked against the set of the parameters indicated by the company representatives, during the case study performed by the authors. The information are collected through the questionnaire designed by the authors. The questionnaire enables a brief insight into the business and organization of distribution in the company. Existing distribution channels are ranked, also, they are ranked together with the suggested one.

2 Methods of Multi Criteria Analysis

In multi criteria decision making, there are different types of multi criteria problems that are described by a suitable mathematical model: Multiple Objective Decision Making (MODM), Multiple Attribute Decision Making (MADM) or Multi Criteria Analysis (MCA).

The MODM model is appropriate for "well-structured" problems. Well-structured problems are those in which the present state and the desired future state (objectives) are known as the way to achieve the desired state.

The multi criteria decision making models (Table 1) are appropriate for "ill-structured" problems. Ill-structured problems are those with very complex objectives, often vaguely formulated, with many uncertainties, while the nature of the observed problem gradually changes during the process of problem-solving. The weak structure makes it impossible to obtain a unique solution. The ambiguity originates from the structure of goals/objectives, which is complex and is expressed in different quantitative and qualitative measurement units.

The selection of distribution channels is a complex problem, therefore this paper is focused on the MCA methods. Some of the MCA methods are explained hereinafter.

The ELECTRE (Elimination Et Choice Translating Reality) method is the first time introduced by Bernard Roy in 1965, meaning elimination and choice of expression of reality. Initially, the ELECTRE method was used to select the best action from given set actions that have evolved and have been applied to three major problems: sorting, ranking and selecting. The advantages of this method are an unlimited number of criteria to rank alternatives and the possibility of quantitative and qualitative expression of the criteria and their importance. The disadvantage of this method is the inability to be applied in circumstances where the decision-maker has not determined the advantage of a certain criteria over others.

The PROMETHEE method (Preference Ranking Organization Method for Enrichment Evaluations) was developed in 1982 by Jean-Pierre Brans, originally for the healthcare problems, but it has also been quickly developed for various applications in banking, medicine, tourism, chemical industry, etc. The biggest advantage of this method is its ease of application. There are two steps to applying the PROMETHEE method: Construct a relation for each criterion in the set of alternatives and using these relationships to solve a multicriteria problem. In the first step, a complex relation of generalized preferences form criterion. Preference indices are

Table 1 Multi criteria decision making models

Criteria for comparison	MODEL	
	Multiple Objective Decision Making (MODM)	Multi Criteria Analysis (MCA)
Criteria defined	Objectives	Attributes
Objective defined	Explicitly	Implicitly
Attributes defined	Implicitly	Explicitly
Constraints	Active	Inactive
Alternatives defined	Implicitly	Explicitly
Number of alternatives	Large	Small
Decision maker's control	Significant	Limited
Application	Design	Choice, evaluation

defined, and this complex preference relation is displayed by the graph of preferences. The essence of this step is that the decision-maker must express his preferences between the two alternatives against each of the criteria. The preference relation constructed in this way is used to calculate output and input flows in the graph for each alternative. Based on these flows, the decision-maker can introduce a partial order into the set of alternatives (PROMETHEE I) or a complete order (PROMETHEE II).

TOPSIS (Technique for Order of Preference by Similarity to Ideal Solution) is a method developed by Hwang and Yoon. This method is a sequencing technique preference for similarity to an ideal solution. It is based on choosing an alternative that should have the greatest geometric distance from the worst solution and the smallest geometric distance from the ideal solution. The premise of this method is that the criteria increase or decrease it uniformly, which leads to the definition of the ideal and the negative ideal solutions. This method compares the importance of the criteria, normalizes the results and computes the geometric distance between the ideal solution and each alternative. By comparing these distances ranking of alternatives is ensured. It is mostly used to rank and improve solution performance, and less often used for the decisions themselves.

The VIKOR method is applied when the decision-maker has no real vision to deal with a problem. It can be said that the VIKOR method creates a compromise between a decision maker's desires and the real options currently available to deal with the problem. The most significant reason for using it in practice is related to a variety of objective circumstances that cannot be achieved by theoretical calculations, and it is needed to find a realistic "closest" optimal solution. A typical example is the construction of roads.

The Analytic Hierarchy Process (AHP) is a method into the most well-known and commonly used methods for multi-criteria decision-making when the process deciding whether to choose one of the available alternatives or ranking them is based on multiple criteria that have different relevance and are expressed using different scales. It is a very important multi-criteria decision-making method that has its application in solving complex problems whose elements are goals, criteria, and alternatives.

The AHP method enables designing an interactive hierarchy of problems that serves as a preparation for decision making, and then pairs of criteria and alternatives are compared, and all of them are eventually synthesized comparisons and determine the weight coefficients of all elements of the hierarchy. It also enables interactive analysis sensitivity. Sensitivity analysis reveals how each change in input is affected outputs to discover why this alternative turned out to be the best and what would happen if the criteria were to change slightly to get a broader picture of the solution to the problem. Sensitivity analyses can simulate the importance of criteria and observe changes in the alternatives rank. The analysis is performed to determine whether the ranking of alternatives is sufficiently stable relative to accept changes to the input, or whether there could be small changes in the input data that lead to major changes in the ranking of alternatives. If by changing the input data by 5% in all possible combinations, no change in the ranking of alternatives appear, the stability of the results is considered to have been achieved.

3 Principles of the Analytic Hierarchy Process (AHP Method)

The Analytic Hierarchy Process (AHP) is one of the most used methods in decision making processes, developed by Thomas Saaty in the early 1980s. It aims to quantify the relative priority of the given set of criteria and alternatives against the appropriate value scale (Saaty 1980, 1990).

The method is based on the mathematical theory of eigenvalues and eigenvectors, by pairwise comparison matrix $W = |w_{ij}|$ $(i, j = 1, 2, \ldots, n)$. Experts compare all the evaluation criteria w_i and w_j (i, j = 1, 2, …, n) where n is the number of the criteria compared, as defined by Eq. 1. The elements of the matrix represent the relationships between the criteria weights (Eq. 2).

$$w_{ij} = \frac{w_i}{w_j} \tag{1}$$

$$W = \left[W_{ij}\right]_{n\times n} = \begin{bmatrix} W_{11} & \cdots & W_{1n} \\ \vdots & \ddots & \vdots \\ W_{1n} & \cdots & W_{nn} \end{bmatrix} \tag{2}$$

The comparison is not complicated as it is easier to compare the criteria in pairs than all at a time. It indicates if one criterion is more significant than the other and to what level the priority belongs. Thomas Saaty suggested a five-point evaluation scale (1-3-5-7-9) to be used for evaluation. The evaluation of the criteria ranges from $w_{ij} = 1$, when w_i and w_j are equally significant, to $w_{ij} = 9$, when the criterion w_i is much more significant than the criterion w_j (Saaty 1980, 1990).

The method starts by structuring a decision-making problem as a hierarchy model in the form of an upside-down tree where the main goal is placed on top. Example of a hierarchy model with two criteria in the first level, corresponding sub-criteria in the second level and four alternatives in the third level is shown in Fig. 1.

First the priorities of the criteria are calculated and then the priorities of the alternatives against each criterion. The elements of the vectors of priorities p can be calculated by means of the Expression (3).

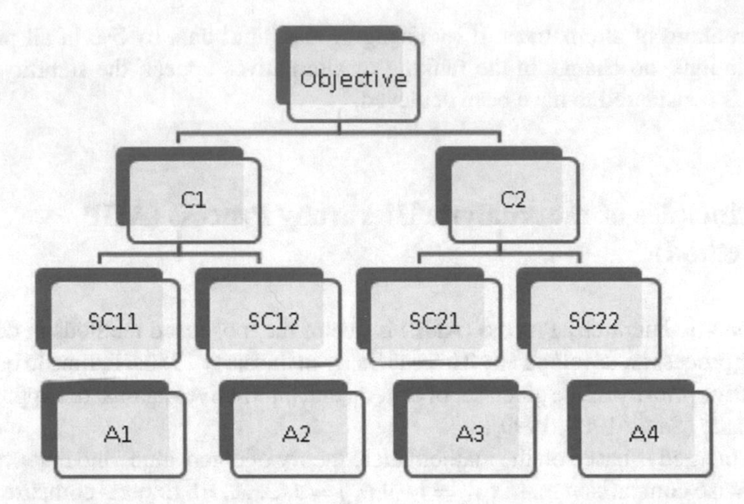

Fig. 1 Hierarchy model

$$p = \begin{bmatrix} p_1 \\ \cdots \\ p_i \\ \cdots \\ p_n \end{bmatrix} \quad \text{where} \quad p_i = \frac{G_i}{\sum_{i=1}^{n} G_i} \tag{3}$$

G_i is the geometric mean of the row i and n is the number of rows of the matrix $W = |w_{ij}|$.

The final rank of the alternatives against all criteria is obtained by grouping the priority vectors of the alternatives in the sublime matrix and multiplying that matrix by the criteria priority vector.

AHP method is very flexible because it produces a simple way to find the relationship between criteria and alternatives. This method and thereby to assess the relevance of the criteria in the real world and determine the interaction between the criteria, in case of complex problems with many criteria and relatively large number of alternatives. By application of this method complex problems could be decomposed in specific hierarchies so the analysis will include quantitative and qualitative aspects of the problem. AHP connects all levels of the hierarchy. This enables the recognition of how the change of one criterion affects to the other criteria and alternatives.

The AHP method allows checking the consistency of the decision maker's estimates after criteria comparing. The necessary condition of the comparison matrix consistency is transitivity of matrix elements' significance. Consistency Ratio CR is calculated as the ratio of Consistency Index (CI) and Random Index (RI). If the CR is less than 0,1 then the decision maker's estimate is consistent, and it is worth investigating why the inconsistency occurred. Some of the advantages and disadvantages of the AHP method are listed below.

Advantages:

- The AHP method integrates both qualitative and quantitative factors in decision making. AHP is a theory of relative measurement in which an absolute scale is used to measure the qualitative and quantitative criteria that are based on expert judgment. Redundancy when comparing two criteria or alternatives leads to the AHP method's low sensitivity to estimation errors.
- Decision-making using the AHP method increases knowledge of the problem and strongly motivates the decision-makers. The decision-making process comes up with an approximate solution of the problem faster than at most meetings and with less expense in the decision-making process. The results obtained can also be used as input for another project or study feasibility, in which much more complex decision must be made.
- It enables the decision-maker to analyze the sensitivity of the results by which checks the stability of the results obtained.
- This method successfully simulates the decision-making process from defining the goal, criteria, and alternatives, by comparing criteria and alternatives in pairs and getting results, that is, prioritizing all alternatives against the goal set.
- AHP successfully identifies and indicates the inconsistency of decision-makers by monitoring inconsistencies in estimates throughout the procedure, by calculating the index and consistency ratio. This is important since decision-makers may be inconsistent in their assessment.
- If used in a group decision-making, this method will greatly improve communication among the group members as they must agree on each criterion and common estimates to be entered into the matrix.
- The decision results in this method contain the ranking of alternatives but also the weight coefficients of the criteria against the target and the sub-criteria to the criteria.
- The great advantage is the existence of quality software tools that support AHP a method such as *Expert Choice* and *Super Decisions*, which make modeling easy and interface are tailored to the average IT educated person.

Disadvantages:

- Great number of pairwise comparisons needed in case of very complex problems.
- Achieving consistency may be difficult.
- When it is difficult to describe the difference in importance among criteria or between alternatives, the Saaty scale may not provide sufficient options.
- The inability to use incomparable alternatives.

4 Specific Features of Confectionery Products Distribution

The importance of the confectionery industry is reflected in the fact that it benefits domestic raw materials, is a significant consumer of packaging materials, directly employs numerous labor force, while indirectly facilitate employment of the population in agriculture, traffic, and commerce. The role of the confectionery industry in the Croatian economy has been emphasized throughout the significant export success of domestic confectionery to the markets of the region and in the world. The most common classification of confectionery products in Croatia is in the following four categories: cocoa products, candy products, chewing gum and flour confectionery products.

Confectionery products belong to the group of food products obtained from three basic raw materials: cocoa fruit, sugar, sugar syrups, and flour. Due to the aforementioned ingredients which are intensively affected by changes in temperature, the appropriate temperature and hygiene regimes should be present during transport and warehousing. The most common causes of spoilage are incorrect transport, handling and sale, as well as unhygienic conditions of production, transportation and handling, increased humidity, pressure, temperature, etc. Therefore, all vehicles used for transportation must be dedicated, clean, closed and resistant to external conditions. Products should be packed in such a way that they cannot be handled by normal handling damage.

In addition to product features, many factors affect the distribution and based on which suitable distribution channels are selected. The criteria are sorted by consumer characteristics, product characteristics, intermediary characteristics, competition characteristics, and managerial and managerial properties. These criteria are analyzed from the perspective of a manufacturing company. To get a more in-depth look at the business and select distribution channels for the confectionery company, the logistics manager of the company was asked to fill in the questionnaire designed by the authors, to collect the relevant data and information. The answers from the questionnaire are summarized below.

5 Evaluation of the Existing Distribution Channels

Technically, the evaluation is performed by using the Expert Choice software, based on the information collected from the confectionary company (Internal data from the confectionary manufacturing company), through the questionnaire designed by the authors. It allows the creation of a hierarchical model of problem-solving in multiple ways and comparing pairs in several ways. In this case, the goal is not to identify the best distribution channel, but to evaluate existing channels for the purpose of ranking.

The main criteria that are relevant for the selection of distribution channels were identified based on the information collected through the questionnaire. This criteria and respective sub-criteria are classified into five categories, as follows:

1. Consumer characteristics;
2. Product features;
3. Intermediaries characteristics;
4. Competition characteristics;
5. Managerial and financial perspective.

With reference to this classification, the criteria and sub-criteria hierarchy model is designed, as shown in Fig. 2.

After the criteria have been assigned respective importance, the calculation of the relative weights is performed, as well as consistency checks for each calculation. As shown in Fig. 3, the highest ranked criterion is *Managerial and financial perspective*. The following criteria are *Intermediaries characteristics* that are also of significant importance, then *Consumer characteristics* with lower importance and finally *Product features* and *Competition characteristics*, which equally affect selecting distribution channels. The inconsistency is 5%, where the consistency ratio less than or equal to 10% is considered acceptable.

The most important sub-criterion in selecting distribution channels from the prospective of *Consumer characteristics* is *Consumer concentration*. The reason for this is obvious: the greater the number of consumers, the greater the profits. So products are to be distributed in densely populated areas, where demand for a particular product is substantial.

Given the *Product features*, the most influencing sub-criterion in selection of distribution channels is *Category management*. Category management or assortment management can be defined as strategic product category management with continuous partnership between sellers and suppliers to maximize sales, profits and meet customer needs.

For *Intermediary characteristics* criteria, the sub-criterion *Price of service* (margin) has the highest relative weight. Margin is a multiple meaning term and can be defined as the difference between the purchase price and the sale price in a store. It is expressed as a percentage of the purchase price.

In terms of *Competition characteristics*, the sub-criterion *Price difference relative to the competition* has the highest relative weight. Each price level leads to a different degree of demand.

Profitability is the sub-criterion that carries the highest relative weight in *Managerial and financial perspective*. Profitability reflects the success of the business, which is what the profit from the sale is. According to the questionnaire, it is also the main criterion (sub-criterion in analysis) according to which distribution channel is selected.

The alternatives in the AHP method are potential solutions. In this case, alternatives are existing distribution channels that are evaluated in pairs, against the individual criterion. There are four default alternatives (existing distribution channels):

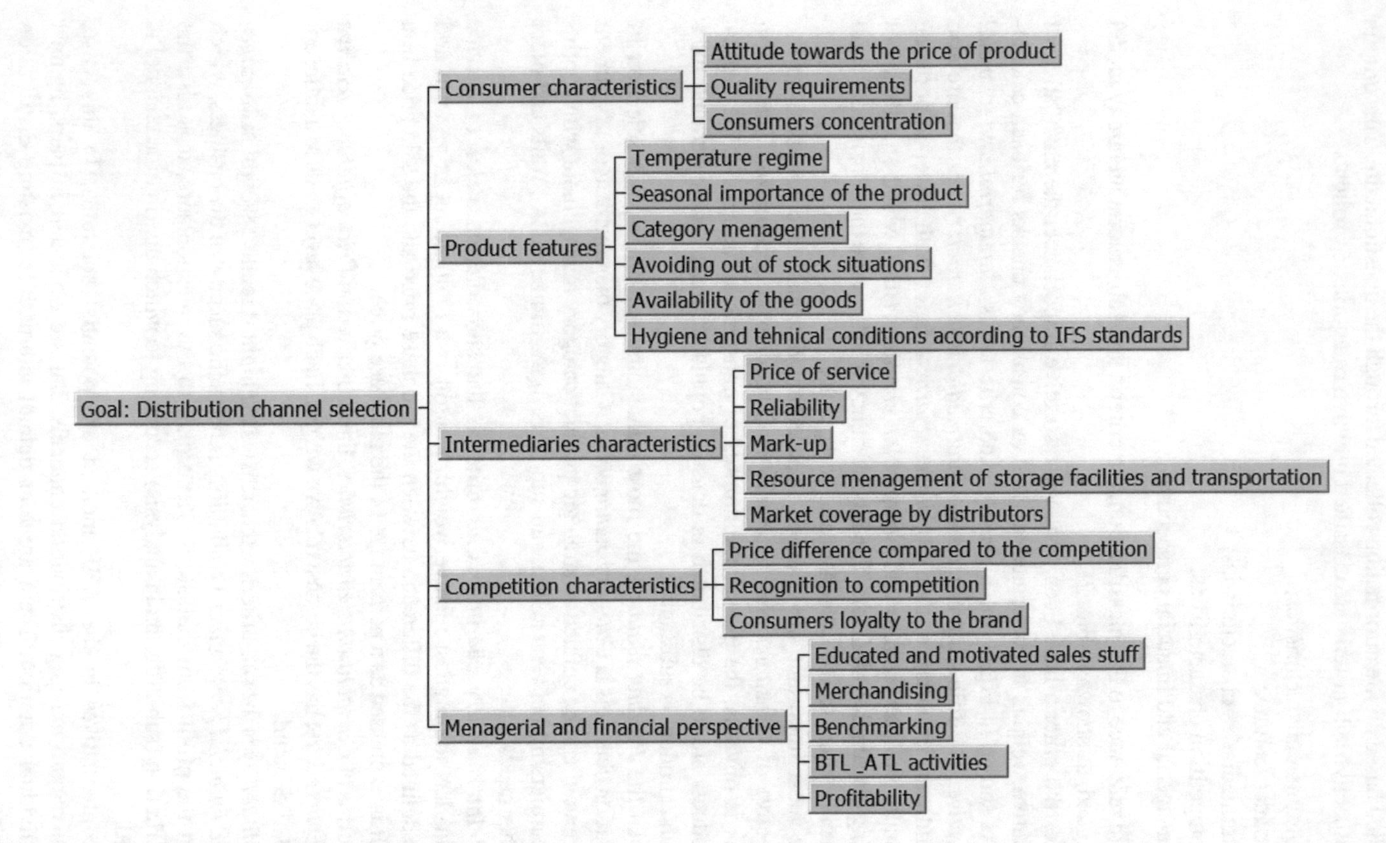

Fig. 2 Hierarchy structure of the criteria in AHP model

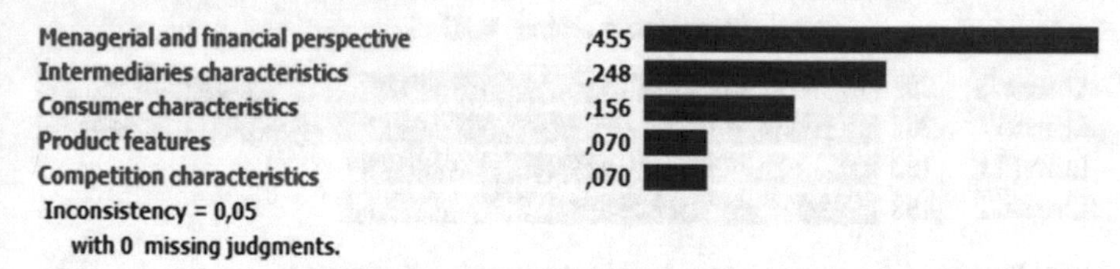

Fig. 3 Relative weights of criteria

1. **Channel 1** (alternative 1) represents domestic wholesale trade (in Croatia). The products are delivered from the factory (primary distribution) to the warehouse of the customer (logistic distribution center), while the customer performs secondary distribution to its other facilities.
2. **Channel 2** (alternative 2) which represents international wholesale trade. The products are exported via the freight forwarders facilities or directly to the foreign customers.
3. **Channel 3** (alternative 3) is retail trade, assuming the goods are shipped directly from the factory to the retail store or from the LDC in Osijek. More specifically, it represents retail in Osijek.
4. **Channel 4** (alternative 4) also represents retail, but the goods are dispatch from LDC Zagreb.

The a.m. alternatives (Channels 1 to 4) are ranked with respect to the hierarchy structure of the criteria in AHP model (Fig. 2), i.e. against each of the five criteria categories:

1. Consumer Characteristics;
2. Product features;
3. Intermediaries characteristics;
4. Competition characteristics;
5. Managerial and financial perspective.

The Graphic presentation of the existing distribution channels rankings are given in Figs. 4, 5, 6, 7 and 8 respectively.

The sales performance of a store can be estimated based on the number of people who pass on an average day, the percentage of those who enter the store, the percentage of those who also buy something and the average income per sale. Confectionery products belong to the group of food products and categorized as FMSG (Fast Moving Consumer Goods). These are relatively low unit prices and most often bought impulsively or unplanned. Such stores must be located near residential areas, where demand for such products is constantly high.

Synthesis with respect to: Consumer characteristics
(Distribution channel sele > Consumer characteristics)
Overall Inconsistency = ,01

Fig. 4 Ranking with respect to *consumer characteristics* criteria category

Synthesis with respect to: Product features
(Distribution channel sele > Product features (L: ,070)
Overall Inconsistency = ,04

Channel 1 ,205
Channel 2 ,186
Channel 3 ,300
Channel 4 ,310

Fig. 5 Ranking with respect to *product features* criteria category

Synthesis with respect to: Intermediaries characteristics
(Distribution channel sele > Intermediaries characteri)
Overall Inconsistency = ,03

Fig. 6 Ranking with respect to *intermediaries characteristics* criteria category

Synthesis with respect to: Competiton characteristics
(Distribution channel sele > Competiton characteristic)
Overall Inconsistency = ,02

Fig. 7 Ranking with respect to *competition characteristics* criteria category

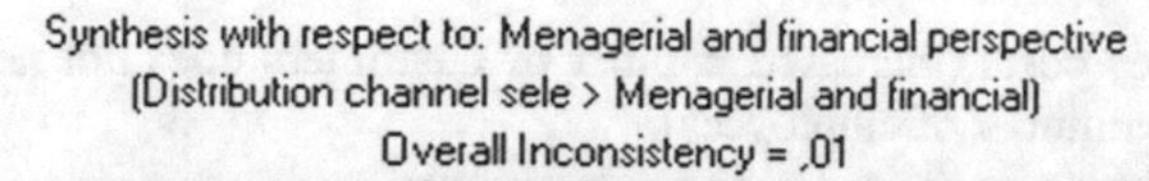

Fig. 8 Ranking with respect to *managerial and financial perspective* criteria category

When selecting a distribution channel, due attention should be paid to the international market, as well as to the possibilities of modifying and adapting existing distribution channels to dynamics of the business specifics and competition. It also has a great impact on wholesale trade on domestic market (in Croatia).

6 Proposal of Introducing Additional Distribution Channel

Removal of customs borders within the single European market has opened a great potential for e-commerce (purchase and sale of goods, services, and information on-line). It takes place between retailers and consumers without physical contact, using one or more means of remote communication via the Internet.

For a company to be successful in using e-commerce as a distribution channel, it must strive to be constantly visible and easy accessible on-line, through a functional and user friendly interface, available at its web site. This will determine the success or failure of the e-commerce distribution channel. The on-line features a business must have for e-commerce are quality of the web site, quality of information and quality of service. The company must provide the system which has high integrity where all client information are safe and secured. Important thing is that the system is designed so that the potential customer can easily manage, understand and use the e-commerce web site, with average computer skills (Koster 2003).

While retailing, the entire e-sales system is customer focused, from the best and most credible product presentation, benefits for customers (discount codes, loyalty rewards, etc.), customer service and product delivery. Since the customer is not physically present at the store to evaluate the product features, the web site must provide sufficient product information and transactions must be constantly monitored by the customer service. An important segment of e-commerce is customer feedback, in order to improve the service quality level, which is crucial for retaining existing customers, as well as to attracting new customers.

The main advantages of e-commerce are:

- constant availability of outlets throughout the year,
- quick access to information,
- global availability,

- lower sales costs (increased number of customers does not necessarily require increased number of employees),
- lower ordering costs (ordering can be automated),
- access to new markets and continuous creation of new business opportunities,
- cheaper marketing that reaches a wider audience,
- ability to monitor customer behavior from their first mouse click to the final purchase,
- providing better customer service,
- reducing customer support costs.

Apart from the advantages of e-commerce, there are also some potential negatives that arise from an increased risk of misuse of information technology. These risks are related to:

- securing data from destruction,
- protecting the confidentiality of certain information contents,
- protecting the privacy of the individuals,
- misuse of customer information by e-commerce company or third person,
- liability for the transactions settlement,
- protection of national interests in case of strategically important products.

Introducing e-commerce as additional distribution channel is proposed, taking into account the effects of a.m. advantages and risks of e-commerce in distributing FMSG. However, the confectionery products belong to the category of food products, so specific conditions in transport, manipulation and warehousing must be complied, as explained in the Chapter 3. This generates higher logistic costs and would only be profitable if quantities of goods are sufficient.

The e-commerce, as additional distribution channel (Channel 5), is now ranked together with the existing distribution channels (Channels 1 to 4) with respect to the hierarchy structure of the criteria in AHP model (Fig. 2), which is adapted with three new sub-criteria:

1. *Assortment range*, added to the *Product features* criteria category;
2. *Lead time*, added to the *Intermediaries characteristics* criteria category;
3. *Return of goods*, added to the *Managerial and financial perspective* criteria category.

In subsequent ranking of all sub-criteria in adapted AHP model, *Assortment range* is second from the bottom in the *Product features* criteria category, while *Category management* still ranks first. Sub-criterion *Lead time* is also second from the bottom in the *Intermediaries characteristics* criteria category. Sub-criterion *Return of goods* has the lowest rank in the *Managerial and financial perspective* criteria category.

Final rankings of all five alternatives (additional distribution channel together with the existing distribution channels) against the adopted hierarchy structure of the criteria, are presented in Figs. 9, 10, 11, 12 and 13 respectively.

Fig. 9 Final ranking with respect to *Consumer characteristics* criteria category

Fig. 10 Final ranking with respect to *Product features* criteria category

Fig. 11 Final ranking with respect to *Intermediaries characteristics* criteria category

Fig. 12 Final ranking with respect to *Competition characteristics* criteria category

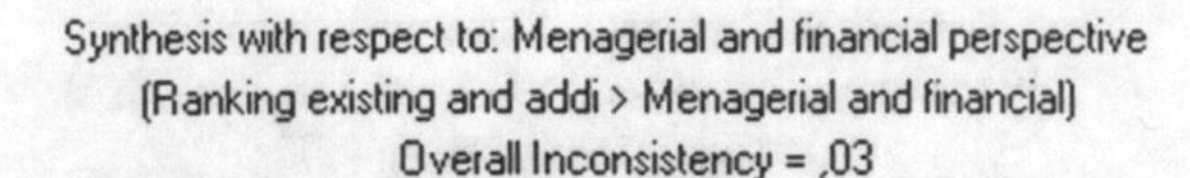

Fig. 13 Final ranking with respect to *Managerial and financial perspective* criteria category

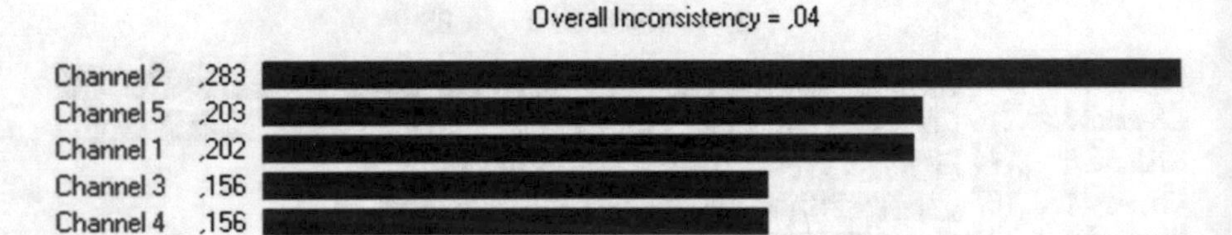

Fig. 14 Total final ranking of the distribution channels priorities

In the total final ranking of all alternatives against all criteria (adapted AHP model), which is presented in Fig. 14, the proposed additional distribution channel (e-commerce: Channel 5) is ranked second from the top, with inconsistency ratio within acceptable limits (<10%).

7 Discussion of the Results

The existing distribution channels are ranked against the set of criteria and respective sub-critera indicated by the company representatives, based on the information collected through the questionnaire, as follows:

- Channel 3 (retail trade, assuming direct shipping from the factory/LDC in Osijek to the retail store) is ranked highest with respect to the *Consumer characteristics* criteria category (Fig. 9).
- Channel 1 (domestic wholesale trade in Croatia, the products are delivered from the factory to the customer's warehouse, while the customer performs secondary distribution to its other facilities) is ranked highest with respect to the *Product features* criteria category (Fig. 10).
- Channel 2 (international wholesale trade, the products are exported via the freight forwarders facilities) is ranked highest with respect to three criteria categories: *Intermediaries characteristics, Competition characteristics* and *Managerial and financial perspective* (Figs. 11, 12 and 13).

- Channel 4 (retail from LDC in Zagreb) is ranked second from the top with respect to the *Consumer characteristics* criteria category (Fig. 9), second from the bottom with respect to the *Iintermediaries characteristics* criteria category (Fig. 11) and the lowest with respect to two criteria categories: *Competition characteristics* and *Managerial and financial perspective* (Figs. 12 and 13).

With reference to introducing e-commerce as additional distribution channel, the hierarchy structure of the criteria in AHP model (Fig. 2) was adapted with three new sub-criteria and the number of alternatives was increased by adding e-commerce as the fifth alternative (Channel 5).

In subsequent total ranking of all five alternatives (additional distribution channel together with the existing distribution channels) against the adopted hierarchy structure of the criteria, (three new sub-criteria were added), the proposed additional distribution channel (e-commerce) was ranked second from the top (Fig. 14).

For this reason, it makes sense the company management may consider introducing the proposed additional distribution channel in its distribution network. For more objective and reliable assessment, a team of marketing and logistics experts should be formed, to conduct extensive analysis, based on more accurate information and detailed data.

8 Conclusion

Proper selection of distribution channels is one of the main assumptions for the effective distribution of products on the market. The application of multi criteria analysis methods enables the systematic evaluation of distribution channels. The results can serve as a basis for determining the optimal set of distribution channels for a given product.

In order to outline the possibility of applying AHP method in evaluation of the distribution channels, the case study on the confectionary manufacturing company was carried out. The evaluation was performed by using the Expert Choice software, based on the information collected from the confectionary company. In this case, the goal was not to identify the best distribution channel, but to evaluate existing channels for the purpose of ranking.

Introducing e-commerce as additional distribution channel was proposed, taking into account advantages and risks in distributing FMCG. With reference to that, the initial hierarchy structure of the criteria in AHP model was adapted with three new sub-criteria and the number of alternatives was increased by adding e-commerce as the fifth alternative.

In subsequent total ranking of additional distribution channel together with the existing distribution channels, against the adopted hierarchy structure of the criteria, the proposed additional distribution channel (e-commerce) was ranked second from the top. For this reason, the company management may consider adopting e-commerce into the company distribution network.

The quality of the results is limited by the information available in the research conducted within the case study outlined in this paper. To obtain a more objective and reliable solution, a team of marketing and logistics experts should be brought together to carry out a more comprehensive analysis of distribution channels, based on more detailed information and insights gathered in each specific case.

References

Internal data from the confectionary manufacturing company (2019)

Matić B (2016) Međunarodno poslovanje-institucije, pravila, strategije. Ekonomski fakultet, Zagreb

Mierzwa D, Zimmer J (19–20 June 2017) The analysis of competitive behavior of enterprises in the chocolate and confectionery industry. Proceedings of the 16th international conference: Tomaszowice, Poland. pp 517–528

RBM De Koster (2003) Distribution strategies for online retailers. Erasmus University Rotterdam

Saaty TL (1980) Analytic hierarchy process. McGraw-Hill, New York

Saaty TL (1990) How to make a decision: the analytic hierarchy process. Eur J Oper Res 48:9–26

Segetlija Z, Mesarić J, Dujak D (2011) Importance of distribution channels—marketing channels—for the national economy, Faculty of Economics in Osijek

Development Barriers of Eurasian Container Transportation

Natalia Lakhmetkina, Alexander Oleinikov, Alexey Pilipchak, and Elizaveta Dmitrieva

Abstract The process of widespread globalization brings international relations to a fundamentally new stage of development. The most significant is transportation in the Eurasian space, where China occupies one of the leading positions in the import and export of products of various nomenclatures. Container transport is particularly important. The object of this study is container transportation of goods through the countries of the Eurasian economic Union (EAEU), the People's Republic of China (PRC) and the European Union (EU). The relevance of this topic is due to the existence of certain barriers and the need to overcome them for the development of cargo turnover in this direction. The theoretical significance of the work consists in a comprehensive study of Euro-Asian cargo routes, identifying the vulnerabilities of existing transport services, and identifying technological and organizational barriers to the development of container transport.

Keywords Container transportation · Development barriers · Eurasian space · Container trains

1 Introduction

The main aim of EAEU as an international integrative unity the draw-up and coordination of the unified policy in all branches of the economy, providing for the free flow of workforce, finance, goods and services. The agreement on creation came into force on January 1, 2015 connecting a number of countries into a generally organized system.

N. Lakhmetkina (✉) · E. Dmitrieva
Russian University of Transport (MIIT), Moscow, Russia
e-mail: naturla@mail.ru

A. Oleinikov
Freight Village RU, Kaluga, Russia

A. Pilipchak
Beijing Jiaotong University, Beijing, Russia

M. Petrović and L. Novačko (eds.), *Transformation of Transportation*, EcoProduction,
https://doi.org/10.1007/978-3-030-66464-0_9

Nowadays the union includes the following countries: Russia, Kazakhstan, Belarus, Armenia and Kyrgyzstan. EAEU membership involved joining the Customs Union, thus applying unified duties and fees. In this aspect we should note the Beijing initiative to create the economic belt of the Silk Route which involves the introduction of common investment to develop trade between EAEU and China.

Speaking about freight statistics from China to Europe no doubt, that over 90% of the market is provided by the sea transport. Its main advantages, that provide in sustainable leadership now, are relatively low cost and great carrying capacity of vessels. Despite the growth of rail's share in Chinese imports to EU comparison with sea transport is still impossible. This is connected with a number of factors.

The goods transported have special parameters which have direct influence on transporting conditions and package choice. Physical state is considered (hard, liquid or gas), as well as properties (perishable, flammable, hazardous). The tonnage of freight is important, however, there are no high tonnage freight in EU–China trade apart from Swedish iron ore, Scandinavian timber and French grain, which opens up opportunities for total containerization of fright flows in this direction. It is also important to take into account the value of freight, which is an effective indicator of profitability of any mode of transport. Interconnection is explained by the share of transportation in the final product cost for the consumer, compared to other economic parameters.

Another factor of distributing freight flow by modes of transport is the geographic location of trade regions. Significant distance between UE and China leads to the priority of sea, rail and air transport, while road transport has many limitations and is used mostly in the border area. For a long time it has been sea transport that provided for the development of logistic network, however active modernization of the infrastructure, improvement and coordination of tariff policy, as well as constant marketing to attract freight on transit transport modes encourages clients to choose rail.

Another important group of factors are technological parameters of transport modes: speed and terms of delivery, regularity of transportation, reliability of proving the safety of freight in transit, door to door delivery.

Analyzing interconnection of tariffs on transportation of container freight, it is evident that low pricing dynamics on sea transport doesn't have almost any influence on demand, which is provided for by keeping the profitable freight and other advantages. Yearly rates of growth and fall have kept at 30–50%, without significant correlation with the growth of demand. Lower prices in no way provided for more sea transportation, and EU–China container turnover has been stable. However, it's important to mention here, that most ports are not initial and terminal points of freight flows, and involve interaction with other modes of transport. It is also necessary to consider delivery and trans-shipment costs. These extra costs provide for a huge share in the final cost of the freight delivery (Statistical Service of the European Union 2020).

The connection of rail container freight and their influence on freight flow are more evident. Commercial and regular transportation started in 2014, with only trial one-time shipments without great economic effect. Dynamic growth of rail container freight volume in the Eurasian space has been observed since the dramatic price reduction on the service.

There is no single "through" rate on the route. That is provided for by the fact that according to the route timetable of the Chinese container operator CRCT, all kinds of freight are transported via EAEU. Each member of the transcontinental route set its own transportation conditions, most often their conditions being different. Moreover, another type of cost is rolling stock using, when rolling stock can belong not to the consigner, but be rented.

Thus for a long time rail freight has been incompatible due to high cost, and a few companies couldn't influence the total cost of the transportation. However, the Chinese government saw some special potential in this area, and offered a temporary solution to increase the attractiveness of rail freight—subsidizing the most significant directions.

At present, China applies measures to increase attractiveness, such as a decentralized subsidizing system, to transport goods in the Eurasian space by rail. Investments are made from the budget of the provinces and cities concerned, and exclusively in relation to international container transport by rail. Cash levels also vary significantly by region. This mainly concerns the cities of Central China (Chongqing, Sichuan, Hubei, Henan) due to the remoteness from the ports and the shorter distance to Europe by land. The average amount of subsidies varies from $ 1500 to $ 7000 per FEU (forty-foot equivalent unit), with the tariff for transportation from China to the EU countries about $ 7500–10,000 (Table 1). This factor allows reducing the total cost of transportation and allowing railways to compete with sea trade routes, and the growth trend in freight traffic continues with positive prospects.

According to various estimates, the total amount of subsidized rail container transportation in the Eurasian space by the Chinese authorities annually is about $ 88 million, which significantly changes the policy regarding the distribution of freight traffic by various modes of transport in this direction. Such a reduction in freight

Table 1 Subsidies for railway transport of China–EU by region (China Railway Corporation 2019)

Subsidizing region	Route	Distance, km	Time, day	Pass-through tariff, $/FEU	Subsidies, $/FEU	Final tariff, $/FEU
Zhengjiang–EU	IU–Madrid (Spain)	13,052	20	10,000	5500	4500
Chongqing–EU	Chongqing–Duisburg (Germany)	11,000	15–17	8000–9000	3500–4000	4750
Henan–EU	Chengzhou–Hamburg (Germany)	10,245	16–18	10,500	3000–7000	5500
Jiangsu–EU	Suzhou–Warsaw (Poland)	11,200	12–15	7500	1500	6000
Sichuan–EU	Chengdu–Lodz (Poland)	9965	12–14	8500–10,290	3000–3500	6150
Hubei–EU	Wuhan–Czech Republic (Poland)	10,700	15–17	12,000	4000–5000	7500

charges expands the commercial appeal for a wide range of goods and stimulates an increase in the frequency of dispatch of goods.

Along with cost factors, the speed and regularity of supply plays an equally important role, where rail container transportation has a significant advantage over sea.

Now the main part of the routes is implemented in regular transportation 2–4 times a week, in some areas more often. Priority is given to deliveries directly linking the transcontinental centers of the EU and China. For example, on the Duisburg (Germany)–Chongqing (China) route, the number of container train departures is 23–24 times a week, on the Chengdu (Sichuan, China)–Lodz (Poland)/Nuremberg (Germany)/Tilburg (Netherlands) route—31–32 times a week. This demonstrates the results of the efforts of Chinese, Russian, Kazakh, European and private railway operators, due to which the frequency of deliveries in this direction took on a regular character and significantly exceeded the figures for sea transportation.

The speed and regularity of shipping in the Eurasian space by sea, taking into account the use of modern container ships, still remains on average 35–45 days. At the same time, the risk of delays due to the influence of climatic and many other factors cannot be ruled out. When assessing, it is worth considering not only the distance "from port to port", but also the time to consolidate cargo in ports (about a week), as well as the further movement of material flows to the destination. Despite the sufficient rhythm of shipping, in this comparative factor, the maritime container shipping market cannot compete with rail.

2 Routes Analysis

When analyzing routes, a significant part of the accelerated container trains of the direction in question passes through Russia. Attempts to create optimal routes began in 2016, when the X8426 China–Europe container train was launched. It proceeded from the Shilong railway station in the city of Dungan to the city of Duisburg (Germany), the border crossing was carried out in Manchuria, and then the train passed through the territory of Russia, Belarus, Poland and Germany. The total travel time was 19 days, which is twice as fast as maritime transport. Many product lines were transported in containers: furniture, household appliances, telecommunications equipment, car parts, etc. The total length of the route was 13,488 km, which is currently one of the record levels for the length of land lines of container trains in the Eurasian space. In the already planned infrastructure development plan for China, the Shilong international railway terminal in Dongan may become one of the largest transport and logistics centers in Asia, which will serve a significant share of rail and multimodal transportation. The only question is the ability of other participants in the transportation process comply with a fairly high standard set (Eurasian Development Bank 2019).

In this context, infrastructure and other negative factors slowed down the implementation of the set plan for the land passage of cargo flows. This outlined the priority goals of creating a set of measures to modernize transport processes. The first accelerated container train FMS-Express eastbound was formed at the Silikatnaya station of the Moscow railway, which arrived in Vladivostok on January 30, 2017. The service is organized as part of the global project "Transsib in 7 days". The technology involves the organization of the movement of accelerated container trains from the departure station to the destination station, eliminates or minimizes stops on the route. Such a scheme can significantly reduce transit time along the Trans-Siberian Railway by increasing the average route speed from 830 km/day to indicators exceeding the mark of 1000 km/day. In addition, electronic document management and preliminary customs declaration of goods were introduced on this route. The successful launch of FMS-Express contributed to the development of foreign trade relations with China, demonstrating Russia's willingness to take measures to improve rail transport (Bessonov 2019).

Currently, accelerated container transportation continues its constant development. Routes and regularity of shipments become more stable. The greatest growth dynamics in the demand for shipments by this method of transportation is noticeable in many cities in China: Suzhou, Nanjing, Guangzhou, Dalian, Wuhan, Yiwu, Chongqing, Chengzhou and others. This trend is explained by the presence in the above areas of a high concentration of production. Table 2 provides a summary of regular services in the Eurasian space based on the data of China Railway Rolling Stock. The standard length of container trains running on the main routes is 57 conventional wagons (41 FEU). Departures of expedited container trains along these routes are carried out taking into account the needs of consignees.

The practice of delivering containers to any destination station in Russia, China, as well as the EAEU and the EU (Fig. 1). This is a great advantage not only over sea mode of transport, but also in comparison with conventional railway shipments. The question is only in the quality implementation and continued development of the existing service.

The transport and logistics company TransContainer JSC made a major contribution to the development of the trade routes of the Eurasian space by expedited container transportation. The routes implemented by the company pass through the territory of Russia, Kazakhstan and Finland. In addition, together with the Russian–Chinese company Swift Multimodal Rus (part of the international group of companies Swift Transport International Logistics Co. Ltd.), a regular container service was created between China and Yekaterinburg from 50 FEU, including telecommunication and construction equipment, power tools, consumer goods and other goods. Containers come from different cities of China to the port of Yingkou, and then to Shenyang (Liaoning Province), where later the composition of the accelerated container train is formed.

Guaranteed transit time is 10–11 days, which was achieved thanks to the organization of transportation through the Manchuria–Zabaykalsk checkpoint (delivery time is 2–3 times less compared to existing routes through the ports of the Far East with further transportation of cargo via the Trans-Siberian Railway). Container service achieves maximum efficiency through the use of containers that arrived in Urumqi loaded from Europe.

Table 2 Accelerated container trains China–Europe–China (Eurasian Development Bank 2019)

№	The regularity of sending	The point of departure	Delivery time, days	Route	Border crossing point	Transit countries
1	1 per week	Zhengzhou North	15	Zhengzhou–Hamburg	Alashankou/Khorgos	Kazakhstan, Russia, Belarus, Poland, Germany
2	2 per week	Yutian			Erlian	Mongolia, Russia, Belarus, Poland, Germany
3	1–7 per week	Chongqing	15	Chongqing–Duisburg	Alashankou/Khorgos/Erlian	Kazakhstan/Mongolia, Russia, Belarus, Poland, Germany
4	1–7 per week	Chengdu North	12–15	Chengdu–Lodz/Nuremberg/Tilburg	Alashankou/Khorgos	Kazakhstan, Russia, Belarus, Poland, Germany, Netherlands
5	2 per week	Jiashan	12–15	Wuhan–Minsk/Hamburg	Manchuria	Russia, Belarus, Poland, Germany
6	1–2 per week		15	Wuhan–Pardubice/Lodz/Hamburg/Duisburg	Alashankou/Khorgos	Kazakhstan, Russia, Belarus, Poland, Czech Republic, Germany
7	1 per week	Hefei East	18	Iu–Madrid	Alashankou	Kazakhstan, Russia, Belarus, Poland, Germany, France, Spain
8			15	Hefei–Hamburg		Kazakhstan, Russia, Belarus, Poland, Germany

(continued)

Table 2 (continued)

№	The regularity of sending	The point of departure	Delivery time, days	Route	Border crossing point	Transit countries
9	2 per month	Changsha	15	Changsha–Hamburg		Kazakhstan/Mongolia, Russia, Belarus, Poland, Germany
10	1 per week	Iu	12	Iu–Minsk	Manchuria	Russia, Belarus
11	1–3 per week	Suzhou West	12	Suzhou–Warsaw		Russia, Belarus, Poland
12	2 per week	Changchun North	13	Changchun–Dresden		Russia, Belarus, Poland, Germany
13	every day	Shenyang	13	Shenyang–Hamburg		
14	3 per week	Harbin South	13	Harbin–Hamburg		
15	1 per week	Xiamen	16	Xiamen–Hamburg	Alashankou	Kazakhstan, Russia, Belarus, Poland, Germany
16	1–2 per week	Erlian	18	Hamburg–Zhengzhou	Erlian	Mongolia, Russia, Belarus, Poland, Germany
17	1–4 per week	Alashankou	18	Hamburg–Zhengzhou Hamburg–Wuhan	Alashankou/Khorgos	Kazakhstan, Russia, Belarus, Poland, Germany
18	2–7 per week			Lodz/Nuremberg/Tilburg–Chengdu	Alashankou/Khorgos	Kazakhstan, Russia, Belarus, Poland, Germany, Netherlands
19	1–2 per week		20	Madrid–IU	Alashankou	Kazakhstan, Russia, Belarus, Poland, Germany, France, Spain

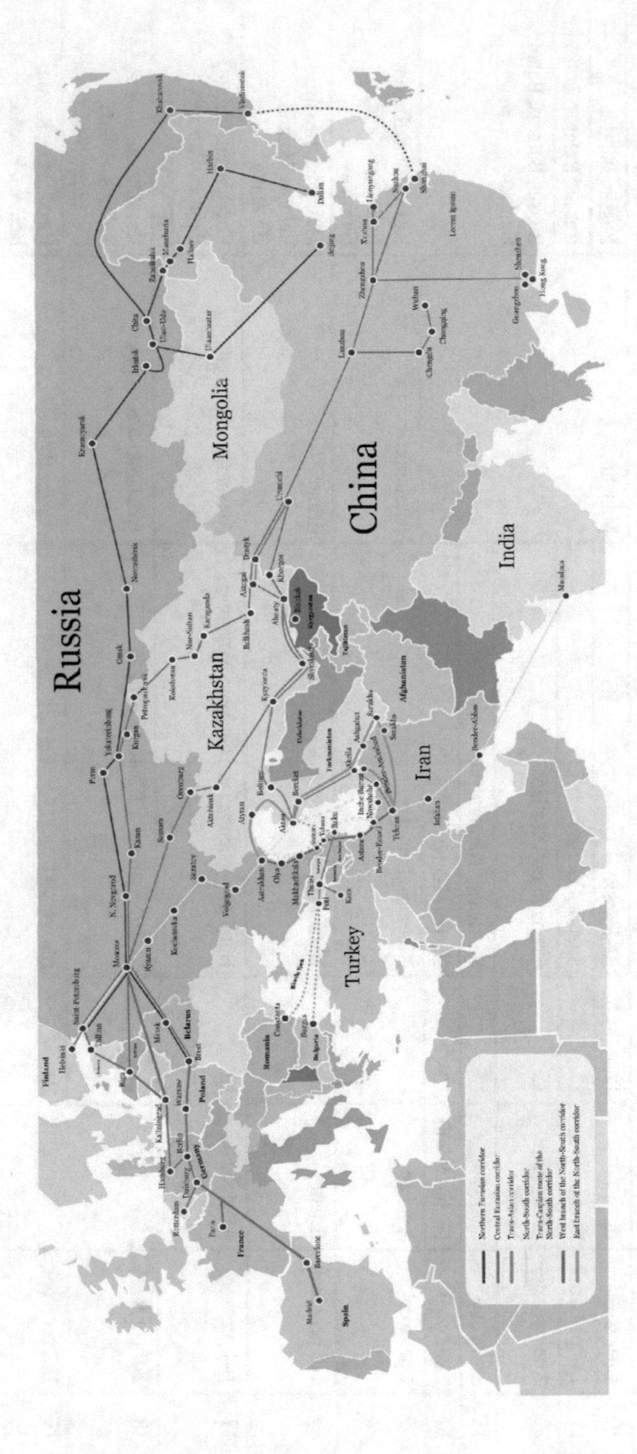

Fig. 1 The route of accelerated container trains in the Eurasian space

An equally important player is United Transport and Logistics Company–Eurasian Railway Alliance JSC (UTLC ERA). Regular shipments from China to Europe are carried out along the following routes: Chongqing–Duisburg, Zhengzhou–Hamburg, Suzhou–Warsaw, Wuhan–Pardubice, Yiwu–Madrid, Chengdu–Lodz. Currently, various departments of the organization are working on the creation and improvement of transit routes.

There are many other private companies on the global market that are implementing accelerated container shipping along the East–West corridor. With a typical query in the Internet search engine, you can find dozens of different offers. Most of them include an individual approach to servicing at all stages of the planning and implementation of transportation, full support of the cargo along the entire route, assistance in customs processes, document management, etc.

The formation of trains takes place in the main commercial and industrial regions of China, and then the finished train follows the destination according to the optimal routes, which allows the client to choose the necessary option on his own depending on the price and delivery time.

3 Development Barriers

The popularity of these shipments is complicated by various development barriers that hinder the growth of traffic volumes. With a gradual increase in the volume and intensity of container flows, the question of the need for technical modernization is becoming more acute and relevant for many participants in the transportation process (Infrastructure barriers). In general, the situation with the throughput as well as the carrying capacities of internal railway networks in the Eurasian space is as follows:

(1) China—even taking into account the promising volumes of imports and exports with the EAEU and the EU, the dynamically developing infrastructure of Chinese railways is able to ensure the satisfaction of the needs of shippers along the route. In the western part of the mainland, high-speed traffic is organized to the city of Urumqi, which, within the framework of one of the many projects for the development of the Central Eurasian Corridor, by 2026 will be able to become a major logistics center in the Urumqi–Dostyk section (Germany).

(2) Russia—as for the main components of the route of the Eurasian transport corridor, infrastructure parameters comply with the standards (fully electrified double-track lines equipped with a self-locking system). It is necessary to continue the modernization of the branch of the Trans-Siberian Railway, which runs to the border with China and Mongolia (sections Karymskaya–Zabaykalsk and Zaudinsky–Naushki). From the point of view of capacity, Zabaikalsk–Borzya and Borzya–Olovyannaya are considered the busiest sections in this direction, and financing has been directed to the reconstruction of these areas. In addition, a comprehensive infrastructure upgrade at Zabaykalsk station is currently being completed.

(3) Kazakhstan—for the development of future transit container cargo flows, a substantial increase in the capacity of the infrastructure in international directions is necessary. This can be achieved through the construction of additional railways on the limiting sections; modernization of signaling and communication devices; creating new routes bypassing major transport hubs; reconstruction of border crossings with China.
(4) Belarus—the bulk of container transit is carried out on the double-track and fully electrified section Osinovka–Orsha–Brest. An important component of the infrastructure is the presence of a fiber-optic cable throughout the transport corridor, which is considered the most modern and efficient information conductor.
(5) EU countries—despite the relatively high level of development of railway infrastructure, the current throughput and transport capabilities of EU member states have limitations and the risk of not fully implementing unhindered transit of the growing container traffic from China.

Another important development barrier of Eurasian container transportation is the lack of a unified network of terminals and logistics centers (TLC) at the moment, which contributes to the achievement of an optimal structure of the direction and distribution of cargo flows based on the needs of each region.

In this regard, the digitalization of transport and logistics processes and the creation of a single integration platform play a key role, because all TLCs have different specializations, sizes, geographical and logistic features. All modes and directions of work of the TLC should be taken into account when building a system of digitalization of processes at the terminals. An important principle of this work is determined by the fact that the terminal intersects the material and information flows of a large number of participants in the logistics process: cargo owners, freight forwarders, line operators, carriers, executive authorities, etc. At the same time, there are both terminal processes themselves and the direct processes of all the other participants. Therefore, the digitalization system of the TLC processes should have both its own management services (loading of the terminal and the temporary storage warehouse, the work of the automobile and railway fronts, loading and unloading equipment), and act as a fairly universal means of integrating processes that are executed and reflected in partner systems or state control systems.

Strategically, in the TLCs digitalization market not IT systems of individual terminals will be promising, but universal systems that are as integrated as possible with carrier systems, state IT systems. Integration with a large number of related systems will inevitably lead to the support of industry standards for exchange and interaction, standardization of exchange services.

As for the Russian Federation, in order to solve this problem, since 2019, the Federal project "Transport and Logistics Centers" has been implemented as part of the Comprehensive Plan for the Modernization and Expansion of the Main Infrastructure, which includes the General Scheme for the development of the TLC network in Russia. However, in order to create a truly effective TLC network, built on the principles of common standards and principles of operation, a special program is needed that is developed and implemented by all countries of the Eurasian space (The Government of the Russian Federation 2019).

Among other states that are actively developing the logistics sector, one can single out Kazakhstan, namely KTZ, which was assigned the status of a national integrated translational operator. Within the framework of the developed plan for the modernization of the country's transport system, various projects are being implemented to integrate infrastructure with the Eurasian international transport corridor. One of the largest is the plan for the creation of the Khorgos–Eastern Gate special economy zone by 2020. The Khorgos International Center for Cross-Border Cooperation, which will also include one of the largest railway ports in the region, will act as a fundamental element. The planned total area of the facility is more than 1600 ha (multimodal logistics zone, industrial zone and infrastructure). In addition, in Kazakhstan there are projects to create TLCs in different city districts: Dostyk, Almaty, Aktau, Astana, Uralsk, Aktobe, and others. Projects are mostly oriented towards ensuring the growth of Kazakhstan's transit potential and the development of trade and transport corridors in the Eurasian space (Goreltsev and Polyakova 2019).

It should be noted that China, despite the currently sufficient throughput capacity of railway lines, continues to actively invest in infrastructure development. Over the year, China intends to increase the number of new projects by 45%, as a result of which it is planned to build 6800 km of the railway network. In the context of accelerated container transportation, it is important to maintain China's interest in the development of the New Silk Road global project. Continuing integration with the transport system of the EAEU and the EU, including the development of border infrastructure.

Development barriers of border infrastructure elements and problems, connected with their passing are to be organized into a separate group (Border crossings). As till 2017 the land China–EAEU–EU freight flow had been low, the capacity of rail infrastructure of most international crossings had been able to cope with the tasks set (Liu et al. 2015).

An increase in the total number of trains crossing the Chinese border with Russia, Kazakhstan and Mongolia in the east, and the Polish border with Belarus in the west, made it possible to reach 3700 trains annually (2400 from China and 1300 to China). For example, according to the 2019 timetable the crossing Dostyk (Kazakhstan)–Alashankou (China) had been passed daily by about 5 container trains. Meanwhile the capacity is for 4 extra train-sets per day. The crossing Zabaikalsk (Russia)–Manchuria (China) is usually passed by no more than 2 container trains, however there is an opportunity to increase the capacity tenfold and it doesn't limit the container transportation. The situation at the Belarus–Poland border is more complicated. Currently, 9/10 routes connecting the EU and China pass through the border crossings of the Belarus–Polish border, where the most intensive traffic is observed at the point Brest (Belarus)–Malaszewicz (Poland). It is expected that after the reconstruction of part of the infrastructure of the Polish railways is completed, border posts with higher capacities will be able to provide transit. On the Belarusian side, projects for the development of this direction have also begun, one of which is the modernization of the Brest–Severny station. A significant increase in the average daily processing capacity coefficient is expected, as well as a reduction in processing time, which will speed up the passage of trains. In addition, to pass the increasing freight flows in the directions West–East, East–West, the infrastructure of the border crossing Bruzki–Kuznitsa Belostotskaya is being developed. However, the Belarus activities to compensate for technological limitations of crossings do not fully solve the problem, as the total capacity is still lower than the planned container flow (Wang and Cullinane 2015; Gajewska and Zimon 2018).

The abovementioned factors though being the limiting ones do not form the main barriers of this group. Speaking about Eurasian routes, the main barrier of border crossings is the different gauge of the railways in Russia, Belarus, Kazakhstan and Mongolia (1520 mm) and European countries and China (1435). The task of reducing downtime of train sets during border and customs operations on the EAEU territory is one of the priorities in the development of container transportation. Due to the introduction on January 1 2018 the Customs Code of the EAEU some progress has been reached: automation of customs systems, refusal to use paper-based documents at declaring, lower declaring time, easier declaring procedure. The use of the new customs code has enabled the integration of the international experience in customs operations. A crucial innovation is only the external border control of the EAEU area, without border controls within the area. For example, container trains on the dense route China–Kazakhstan–Russia–Belarus–Poland are checked only on the borders Kazakhstan–China and Belarus–Poland (if all the required conditions are carried out). Other customs operations are completed depending on the stations' capacities.

Another development barrier of Eurasian container transportation is the group of technological barriers.

(1) The length of the train-set, which is formed by each railway company under the influence of certain factors: the length of the station tracks, train weight, traction capability, a compiled route, technical feasibility of hauls, shunting conditions at the station, etc. Depending on the length of the composed train, a loading plan is formed, in this case, the number of fitting platforms for containers is calculated. The overwhelming majority of Chinese trains approach the border with Kazakhstan with a length of 54 wagons (about 756 m), excluding the locomotive and the space for setting the train. On Russian railways, trains average 71 wagons (about 994 m). This is due to the useful length of the pick-up tracks (850, 1050, 1250 and 1550 m). At the same time, trains from empty wagons are formed even longer—up to 100 wagons. This is due to the weight of the train and the traction power of the locomotive, heavy trains are often shorter than the standard length. In Belarus, the length of the wagon part varies from 57 to 65 (up to 910 m), depending on the sections. In Poland, indicators are much lower and in accordance with regulatory enactments do not exceed 600 m (Lakhmetkina et al. 2018).

The norm of the length of the composition is an important technological parameter, on which the management of operational work strongly depends. For example, from the point of view of the plan for the formation of trains, it is considered more advantageous to reduce the train composition, since this reduces the time for the accumulation of trains and processing of cars, which leads to more favorable economic indicators. However, from the point of view of the motion schedule, an increase in the norm of the length of the composition brings a positive effect, since the size of the movement decreases. Thus, at present, we can talk about the inconsistency of standards in different member countries of the Eurasian routes. If a container train of 65 wagons follows in the direction of Poland, then after reloading containers in Brest, there is a need to form a new train of 43 wagons, while the remaining 22 wagons will be waiting for attachment to the next container train. This invariably entails the emergence of new financial and time costs (Pavlenko and Velykodnyi 2017).

Logistic operators are considering various solutions to the situation. One of them is the UTLC ERA project called UTLK XL Train (Extra Long train). Together with Russian Railways, KTZ, and Belarusian Railways, we analyzed the multiple conditions of increasing demand for transportation, and then developed the technology for sending long container trains to optimize the transportation process, reduce costs, as well as increase the efficiency of using the transit route through the territory of Russia, Kazakhstan and Belarus. Today, the technology is being successfully implemented in the U West service in the Dostyk–Brest section. The trains of 80 conventional wagons are formed, which, when fully loaded, allows the transport of 88 FEU. The average speed of the abovementioned section reaches 1100 km/day. The formation of long container trains is a great opportunity to significantly improve the efficiency and competitiveness of transit logistics services. Currently, the technology is not duplicated on all UTLC ERA routes in the direction of China–EU, but this can be achieved in the foreseeable future, which gives a positive example for other transport operators organizing similar transportation.

(2) Barriers in electrification, which are associated with the use of direct or alternating current, as a result of which there are differences in the voltage of contact networks. In Poland, the Czech Republic and Slovakia, a DC voltage of 3 kV is used. Germany uses similar networks with 3 kV, as well as 15 kV AC. In Belarus, Lithuania and Ukraine—networks with 3 kV DC and 25 kV AC. Given the greatest economic attractiveness when using 25 kV AC, Russia settled on this option and is actively developing it. Realized traction allows you to provide heavier trains with lower cost per unit load. However, it should be borne in mind that when switching from direct current to alternating current, it is necessary to replace the locomotive, as well as re-analyze the train, thereby again increasing costs and reducing the competitiveness of this service (Hrynchak 2019).
(3) The weight of a container train, the norms of which are limited and differ depending on the infrastructure capabilities on individual components of the China–EU route. From the point of view of the transportation process, this forms a barrier to the growth of competitiveness of the railway communication (Fesovets et al. 2019). Thus, technical requirements are being developed for the commissioning of 80-foot fitting platforms, which allow loading of 2 FEU, but the existing conditions do not allow the full launch of the project. Models of two-tier container platforms are also considered, but their use is connected with

the issues of loads on the contact network, as well as with the dimensions of the rolling stock and the approximation of the structure.

(4) The speed of container trains. Many supporters of the development of the transport industry are thinking about how to further increase speed and reduce already attractive delivery times. In practice, however, everything again rests on the existing restrictions.

When comparing speed indicators of all transport participants, the highest values of the set speed are demonstrated by Russia. According to Russian Railways, the average local speed of freight trains is about 42.8 km/h, the average technical speed of freight trains is 47.4 km/h, and the route speed of freight trains is 632.8 km/day. Despite the developed network of Chinese railways, the average train speed in the European direction is about 35.6 km/h, and when approaching the border with Kazakhstan it even drops to 30 km/h, which is due to single-track traffic on the site. As for the EU countries, the average speed of freight trains is even lower—about 18.2 km/h. As a result, the sufficiently fast train traffic through the territory of the EAEU countries is sharply slowed down when entering the EU (Kireev et al. 2018).

Another constraining development barrier is the imbalance of freight traffic (Reverse loading). Despite the rapid growth of container traffic on international routes on the China–EAEU–EU axis, there is no clear interaction between the operators, which more justifies the emergence of a large number of empty containers in this direction. Instead of working more closely to increase competitiveness, carriers continue to compete with each other, without taking into account the factors that set the desired development vector. The essence of the problem is that on routes to the EU, most of the goods from China are transported in containers, but in the opposite direction, the indicators have not yet reached such high rates. Given the high tariffs for warehousing in European unloading areas (0.5–2 €/day per container depending on the terminal), it is economically more expedient to carry out the return shipment when empty (Shestopalova and Zimin 2019).

To search for possible options for filling containers from Europe, a market is being studied and standard innovative technologies are being developed to modernize transportation technology. A promising idea is to create a folding container. After the completion of transportation and subsequent unloading of goods, in the absence of the need for use, it can be folded into a more compact form. The design resembles a standard cardboard box. Six of these containers take the place of one 40-foot, which significantly reduces shipping costs. Another very promising direction is the switching of goods previously transported in a different container.

In addition, continuous monitoring of cargo flow is carried out to balance load and plan new routes. Among them there is an accelerated container train “Milan–Chengdu”. The train passes a distance of 11,000 km through Germany, Poland, the EAEU countries with a destination in China in 14 days. Among the product groups there is Italian furniture, ceramic tiles and industrial machines. Thus, Milan became one of the European cities associated with a large industrial center in China. Moreover, the trains are planned to be filled as cargo is accumulated for shipment; exact schedules have not been formed.

When analyzing development barriers of Eurasian container transportation, one should also take into account the peculiarities of information support for transportation. Now transport organizations invest to the development of innovative information technologies. However, these processes are rather slow, information system of transport process participants requires a standardization and unification. One of the most effective organizational decisions was the introduction of electronic document management and a system for the unified processing of consignment notes in the EAEU countries. The first step was the introduction of mandatory preliminary information and the exchange of electronic documents during customs operations. This measure allowed to significantly reduce the time for cargo clearance at checkpoints, and also demonstrated the potential for possible modernization of information support for all participants in trade and transport relations.

Considering development barriers of Eurasian container transportation mentioned above, the next step is to structure them into 5 groups as a diagram (Fig. 2).

Table 3 gives the calculation of percent share of the importance of different factors depending on the given value from 1 to 10. The data were obtained on analyzing expert opinions of 50 companies, participating in the international transport and logistics event Transport week 2019. The ranging given can help to identify the most vulnerable zones of developing speed container transportation in the Eurasian area for the further search of solutions.

As the obtained result shows, the most significant group of barriers is Border crossings. So the priorities are to be set so that either to find complex solutions to border crossing problems or to find alternative routes of freight transportation. This will be the subject of the following research by the authors.

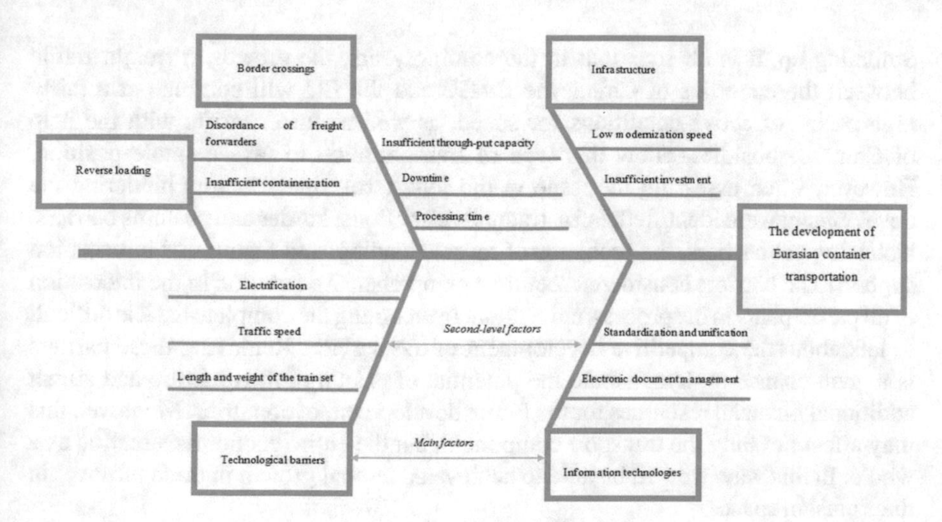

Fig. 2 Development barriers of Eurasian container transportation

Table 3 General range of factors

Factor	Score	Share of total (%)
Infrastructure	–	17.39
Insufficient investment	7	
Low modernization speed	5	
Border crossings	–	34.78
Insufficient through-put capacity	8	
Downtimes	9	
Processing time	7	
Reverse loading	–	18.84
Discordance of freight forwarders	6	
Insufficient containerization	7	
Technological barriers	–	15.94
Electrification	4	
Traffic speed	3	
Length and weight of the train set	4	
Information technologies	–	13.04
Standardization and unification	5	
Electronic document management	4	
$\sum$	69	100

4 Conclusion

Summing up, it is obvious that in the coming years, the growth in freight traffic between the countries of China, the EAEU and the EU will continue at a fairly high pace. Attractive conditions for speed, as well as lower tariffs with the help of Chinese subsidies, allow this type of transportation to take a stable position. However, when assessing the issue in the long term, many factors hindering the development were identified: infrastructure restrictions, border and customs barriers, technological barriers, the problems of reverse loading, and features of information support. The barriers considered require a comprehensive solution in the interaction of all participants in the process and without overcoming the complete list it is difficult to talk about the competitive development of this service. Removing these barriers is a great chance to demonstrate the potential of existing infrastructure and attract additional financial resources for the future development of countries. Moreover, this may affect not only the transport component, but the entire economic situation as a whole. In this way, we will be able to achieve additional growth in trade turnover in the Eurasian space.

References

Bessonov G (2019) Annual trans-Siberian transportation digest. In: Coordination council on trans-Siberian transportation international association (CCTT), Moscow

China Railway Corporation (2019) Railway intermodal transport. China. https://www.china-railway.com.cn/gjhz/gjly/. Accessed 2 Dec 2020

Eurasian Development Bank (2019) Transport corridors of the Silk Road: potential growth of cargo flows through the EEU. Russia. https://eabr.org/analytics/integration-research/cii-reports/. Accessed 10 Oct 2019

Fesovets O, Strelko O, Berdnychenko Y, Isaienko S, Pylypchuk O (2019) Container transportation by rail transport within the context of Ukraine's European integration. In: Proceedings of the 23rd international scientific conference transport means 2019—sustainability: research and solutions, Palanga, 2–4 Oct 2019. Kaunas University of Technology, pp 381–386

Gajewska T, Zimon D (2018) Study of the logistics factors that Influence the development of e-commerce services in the customer's opinion. Arch Transp 45(1):25–34. https://doi.org/10.5604/01.3001.0012.0939

Goreltsev S, Polyakova M (2019) Prospects for the development of terminal and logistics infrastructure in the 1520 space. Institute of Economics and Transport Development. https://iert.com.ru/images/Gorelcev.pdf. Accessed 12 Apr 2020

Hrynchak N (2019) Statistical monitoring of development of the international container shipping market. Bus Inform 12:248–254

Kireev A, Kozhemyaka N, Kononov G (2018) High-speed container transport system. Transp Syst Technol 2:5–18

Lakhmetkina N, Shchelkunova I, Fomicheva O (2018) Logistics solutions for the interaction of modes of transport. World Transp 2:178–187

Liu H-B, Wang X-F, Sun H-S, Zhang W-Y (2015) A study on operational model of container multimodal transport virtual enterprise based on multi-agent technology. Adv Mater Res 1065–1069:3310–3313. https://doi.org/10.4028/www.scientific.net/AMR.1065-1069.3310

Pavlenko O, Velykodnyi D (2017) The choice of rational technology of delivery of grain cargoes in the containers in the international traffic. Int J Traffic Transp Eng 7(2):164–176

Shestopalova D, Zimin A (2019) Improving the efficiency of container transportation along international transport corridors. In: Scientific, technical and economic cooperation of the Asia-Pacific countries in the XXI century, vol 1, pp 217–222

Statistical Service of the European Union (2020) Dynamics of mutual trade of the EU, EEU and PRC countries. European Union. https://ec.europa.eu/eurostat. Accessed 10 Apr 2020

The Government of the Russian Federation (2019) Long-term development program of JSC Russian railways until 2025. Order No. 466. Report of March 19 (2019) Russia. https://doc.rzd.ru/doc/public/ru?STRUCTURE_ID=704&layer_id=5104&id=7017. Accessed 20 Oct 2019

Wang T, Cullinane K (2015) The efficiency of European container terminals and implications for supply chain management. In: Palgrave readers in economics, pp 253–272

Airline Fleet Rotables Staggered Replacement Scheduling Using Dynamic Approach

Miroslav Šegvić, Anita Domitrović, Ernest Bazijanac, and Edouard Ivanjko

Abstract The purpose of this paper is to describe the application of a dynamic approach for scheduling of airline rotable components preventive maintenance in order to minimize earliness costs and maximize on wing time of the components. First, analysis of required service level is performed to determine minimum stock quantity for unscheduled replacements, or removals for repair. Then, a schedule for staggered replacement of propeller blades and hubs for overhaul on a regional turbo-prop aircraft fleet is given, with limited availability of spare blades and other constraints. Results show that a minimal cost solution without part shortages can be obtained.

Keywords Rotable components · Replacement schedule · Staggering · Dynamic approach · Backwards allocation · Earliness costs

1 Introduction

1.1 Airline Rotable Component Maintenance

According to IATA's "Airline maintenance cost executive commentary", global Maintenance, Repair and Overhaul (MRO) spent in 2018 accounted for 9% of total airline operational costs. While engine maintenance is the highest cost segment with 42% of maintenance costs, component maintenance follows with 20% of maintenance costs (IATA's Maintenance Cost Technical Group 2019).

Aircraft maintenance generally includes the tasks specified in Aircraft Maintenance Program (AMP) required to restore or maintain the aircraft's systems, components, and structures in an airworthy condition (Ackert 2010). It is necessary to

M. Šegvić
Croatia Airlines, Bani 75b, Buzin, 10010 Zagreb, Croatia

A. Domitrović (✉) · E. Bazijanac · E. Ivanjko
Faculty of Transport and Traffic Sciences, University of Zagreb, Vukelićeva 4, 10000 Zagreb, Croatia
e-mail: anita.domitrovic@fpz.unizg.hr

M. Petrović and L. Novačko (eds.), *Transformation of Transportation*, EcoProduction,
https://doi.org/10.1007/978-3-030-66464-0_10

maintain the aircraft to satisfy airworthiness requirements, retain the value of the aircraft throughout the life and most importantly to keep it in serviceable condition to create revenue.

Components must also be maintained according to the AMP to remain serviceable and suitable for installation on an operator aircraft. Maintenance and overhaul of components must be completed by strict deadlines. Although preventive maintenance and overhaul reduces failure rates of components, it can do so only to a certain degree. Components still fail in operation at failure rates according to their designed reliability and then require unscheduled corrective maintenance or repair. To avoid interruptions in flight schedule, airlines use the maintenance by replacement concept. The component is being maintained in a workshop while the aircraft returns to operation utilizing replacement or spare component.

Rotable components are removed from the aircraft to perform maintenance, repair or overhaul in component workshop, and are later returned to operator stock as a spare part for future replacements (see Fig. 1). Rotable components represent 95% of total component related maintenance investment (IATA's Maintenance Cost Technical Group 2019), and as such require special attention by the airline operator. For high value components airlines may use rotable inventory. The range and depth of spare components to be kept in store for replacement is determined by an airline during fleet acquisition and initial provisioning process (IATA 2015).

The parameters used for spare parts stock quantity calculation are the number of aircraft in a fleet, number of components per aircraft, Mean Time Between Failure (MTBF), and Turn Around Time (TAT) for the component or the workshop availability. Service Level (SL) describes probability of a spare part being available on stock in a serviceable state, and this parameter is used additionally for provisioning

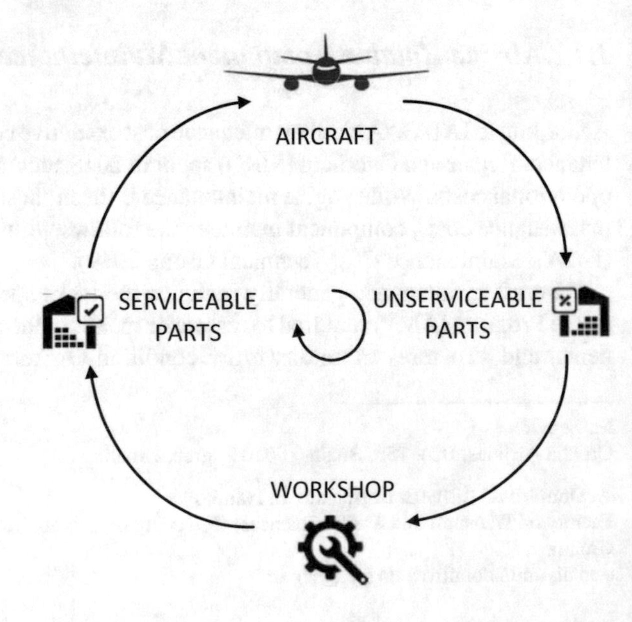

Fig. 1 Rotable component graphical explanation

according to the service level requirement. Throughout the life of the aircraft a revised provisioning is often required to assess surplus or deficiencies made during initial provisioning.

Operators usually acquire aircraft within a narrow time frame. If these aircraft operate in similar conditions, preventative maintenance of major components falls roughly at a same time. When the time comes, the operator has only two options if he wants to keep all the aircraft in operation while components are being maintained. He can increase the number of spare parts to perform group replacements, which is usually an unacceptable costly option, or he can advance the replacements on some aircraft in a fleet to meet the deadlines. If the spare parts investments are limited, it becomes necessary to stagger the fleet component replacements. Schedule for staggered replacement should be made. Scheduling here is not considered as maintenance program development, rather as shop visit or overhaul scheduling for multiple components found on same fleet of aircraft with limited spare parts capacity (Dinis et al. 2019). The goal of such scheduling is to minimize loan, exchange or earliness costs for the airline by providing optimal schedule for components replacement while maximizing on wing time of the components. Following limitations need to be considered for scheduling:

- Operator usually has limited number of expensive rotables available in store for replacement;
- Workshop TAT is determined by the workshop capacity and cannot be affected by the operator;
- Contracted MRO organizations occasionally exceed agreed deadlines, depending on airlines demand and shop capacity.

It can be seen that rotable stores and component workshop capacities present major limitations to rotable component replacement scheduling problem. In order to minimize maintenance costs, operator may group replacements when the deadline for replacement of multiple components on the same aircraft falls in an acceptable timeframe. Man-hours for preparation and disassembly for access can be saved, line occupation reduced, aircraft down time reduced, and shipping costs combined when the same type components are sent to one contracted MRO for maintenance. However some of the remaining resource of the components is lost by reducing the on wing time, and must be accounted for. Failure to plan replacements leads to Aircraft On Ground (AOG) situations, creating large additional costs as exchange or loan fees and revenue loss.

The following section address the problem of making a two-year schedule for preventive replacement of propeller blades and hubs on a regional turbo-prop aircraft fleet using a dynamic approach. Firstly, the closed loop rotable process is described. Then, empirical removal for repair or by defect data is used to determine minimum stock for component. Finally, the schedule is made using backward allocation algorithm, and resource loss due to early replacement is calculated.

1.2 Recent Literature Review

Historically notable work was done in 1998 regarding Heuristics for Multicomponent Joint Replacement through maintenance policies (Hopp and Kuo 1998), paper on Optimal Replacement Policy of Jet Engine Modules from the Aircarrier's Point of View in 2008 (Domitrović et al. 2008), and another paper Dynamic preventive maintenance scheduling of the modules of fighter aircraft based on random effects regression model published in 2010 (Sohn and Yoon 2010).

Only two recently published papers can be found regarding airline scheduling of component maintenance from year 2015. Erkoc and Ertogral present an integer programming model that minimizes the total earliness under rotable inventory and process capacity constraints in an airline, for which the optimal solution can be directly obtained from the Linear Programming (LP) relaxation of the model (Erkoc and Ertogral 2016). Authors also propose a practical exact solution algorithm based on a full-delay scheduling for the problem. According to paper, algorithm can be easily implemented without the need of using any mathematical solver or computer software. A case study is given with sensitivity analysis of the effect of capacity and rotable inventory on the optimal exchange and overhaul processing schedule. The model involves a constant processing time for all rotables. This imposes a significant limitation on the model since constant TAT is never the case in real life scenario. Authors claim that constant TAT can be regarded as statistically correct approximation for presented case.

Second recent paper about component maintenance scheduling is published by Hosseini, Kalam, Barker and Ramirez-Marquez. The paper is dealing with optimization of maintenance schedule to minimize system downtime by performing joint replacement of components whenever joint replacement is economical (Hosseini et al. 2019). Whether it is economical to adjust schedule and perform joint replacement instead of individual replacements, is determined through penalties for earliness and tardiness of removal and for system downtime. Model is solved with a proposed iterative Greedy Heuristic with Local Search Algorithm (GHLSA) through three phase process. The schedule construction, improvement phase producing joint replacement sets, and finally local search phase to balance the penalties. Comparison of effectiveness of GHLSA solution to existing solution methods the Genetic Algorithm (GA) and Simulated Annealing (SA) was performed. Authors claim GHLSA outperforms GA and SA. Sensitivity analysis was performed and showed that selection of optimal time span parameter notably impacts generation of a good initial solution. Multiple approximations were introduced but the authors logically deduced for each that all have same effect for individual and join replacements and therefore do not affect the results. The authors propose further exploration of model usefulness for maximizing reliability instead of minimizing downtime. It should be noted here that although it was never mentioned in the paper that it is related to Airline MRO business specifically, the constraints, references, and general MRO process logic within the paper suggests so. Methodology is entirely applicable to Airline MRO.

2 Use Case Analysis

For the use case analysis in this paper a real-life example and data gathered from one regional airline are used. The operator fleet consist of six twin engine turbo-prop aircraft. Each engine powers a propeller hub with six propeller blades (see Fig. 2).

The number of hubs in the fleet is relatively small, however hub replacement should be grouped opportunistically with blades due for replacement shortly after hub replacement to minimize maintenance cost and aircraft down time. Totally there are 72 blades and 12 hubs installed on the aircraft that comprise the fleet. Operator has 11 spare blades and 1 spare hub in its inventory that can be used for scheduled and unscheduled replacements. The blades and hubs are hard time limited by 10,000 Flight Hours (FH) scheduled maintenance task for restoration (overhaul).

Once the blade or a hub is removed from the aircraft it is sent to an outside workshop for overhaul or repair. When the maintenance on the component is finished, it is being returned and stored on operator stock for future replacements. The general closed loop rotable component maintenance process described can be applied to any repairable component and it is shown in Fig. 3.

It is not uncommon for just only one blade to be removed for workshop maintenance. To reduce the aircraft maintenance costs it is desirable to replace blades that will be due for overhaul on the same aircraft position within a narrow time frame

Fig. 2 Propeller hub with six propeller blades

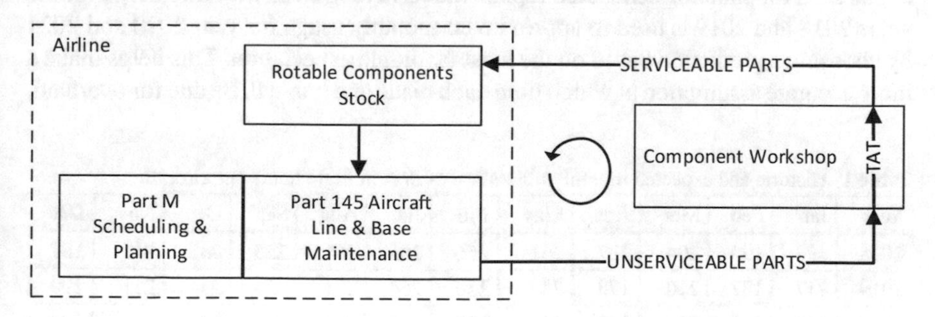

Fig. 3 The process of closed loop maintenance for rotable components

at the same occasion, specifically when the first blade is running out of resource. Analogy can be drawn for hubs. If a hub is scheduled for replacement it is desirable to replace blades installed on the same propeller that are reaching the end of resource on the same occasion. It is up to the operator to decide how narrow this time frame should be to consider joint or group replacement economical. The thirty days rule will be applied in this research. For example, if two blades on the same propeller are to be removed within the thirty days period, they should be removed together. Similarly, if a hub is being replaced and some of the blades installed on the same propeller are to be replaced within a thirty days period, they should be replaced together. Intervals for overhaul cannot be exceeded, so the hub and blades should be removed before any of them reaches 10,000 FH of Time Since Overhaul (TSO).

From the operator experience, the shop TAT for the blades and hubs is between 55 and 65 days depending on demand and season of the year. The cost of the overhaul for the blade is 12,000 Money Units, and the cost of the overhaul for the hub is 26,000 Money Units. With this information one can easily calculate that for each flight hour the blade is removed from the aircraft earlier than needed, the operator loses 1.20 Money Unit, and for each flight hour the hub is removed from service prior to expiration of overhaul interval, the operator losses 2.60 Money Units.

This is relevant for calculation of losses due to staggered replacement. It should be noted that if the component is sent for repair only, it is returned with the same remaining resources, meaning TSO is not reset to zero. It is up to the operator to determine at which TSO, the blade or hub that was removed for repair should be sent for overhaul. However, this is not the subject of this research, and assumption is taken that all blades that are removed for repair are not the ones that are scheduled for replacement within the next two years. That way the unscheduled removals do not affect the schedule. In reality, some of the blades that are scheduled for replacement will be removed earlier by defect and replaced with overhauled ones, which moves them another 10,000 FH down the schedule. The operator has a 2600 FH annual utilization per aircraft. The monthly utilization of the aircraft is intensified during the flying season as shown in Table 1.

The aircraft that were undergoing base maintenance were not included in averages calculated for this table since plan for aircraft base checks can vary slightly and should not be used for planning scheduled replacements in long term. Monthly usage for the years 2018 and 2019 is used to approximate monthly usage for year 2020 and 2021 by choosing a prediction based on the least favorable expectation. This helps make a more accurate assumption at which time each blade and hub will be due for overhaul,

Table 1 Historic and expected monthly utilization of fleet in flight hours per aircraft

Year	Jan	Feb	Mar	Apr	May	Jun	Jul	Aug	Sep	Oct	Nov	Dec
2018	198	193	204	222	261	256	265	266	253	238	211	187
2019	217	187	210	178	245	250	264	261	254	234	211	190
2020	217	193	210	222	261	256	265	266	254	238	211	190
2021	217	193	210	222	261	256	265	266	254	238	211	190

and assure the safety margin. The aircraft enter the hangar for base maintenance annually, during which they do not fly and monthly usage reduces to somewhere between 150 and 0 FH per month, depending on the duration of the maintenance event or scope of work to be done on the aircraft. As previously stated, due to probability of base maintenance schedule perturbations, this will not be accounted for in this research. Since there is seasonality in operating conditions, and a large variation in shop TAT for the hubs and blades, least favorable data should be used for setting of the schedule leaving a safety margin in the plan. The chosen horizon for the schedule is two years, which will be broken down into one-month intervals. Actual plan should be broken down into one-week intervals, but for simplification, one-month interval will be used in this research. The blades and hubs that will be overhauled in that period will be due for next overhaul in roughly four years considering average fleet utilization, surely falling out of horizon of this schedule. The blades and hubs that are due for replacement for each month within the next two years are shown in Table 2. The rows mark aircraft and their propeller positions. The columns denote month of the year in a plan. If one or more blades are due for replacement in certain month on a propeller position, the cell is entered with the number representing the quantity of blades needing replacement. If a hub is due for replacement on a propeller position at a certain month, the respective field denoting the position in entered with the letter "H" following number of blades due for replacement on same position.

3 Service Level

A calculation for spare parts cover is performed when a fleet is first commissioned. It is referred as the Initial Provisioning, and repeated over the lifetime of the fleet as failure rates change (Rutledge 1997). The Airbus Initial Provisioning methodology starts with demand calculation formula:

$$E = fh * n * N * \frac{1}{MTBUR * 365} * TAT, \tag{1}$$

where the demand is denoted as E, aircraft utilization per year is denoted as fh, number of components per aircraft is expressed as n, number of aircraft in fleet as N, and TAT denotes Turn Around Time for the component. Mean Time Between Unscheduled Removal (MTBUR) can be calculated as:

$$MTBUR = \frac{fh * n * N}{Removals}, \tag{2}$$

where Removals denotes the total number of unscheduled removals for the component per year. This formula leads to the following simplified initial provisioning formula:

Table 2 Number of hubs and blades that are due for replacement on each position

AC	Year 2020												Year 2021											
	1	2	3	4	5	6	7	8	9	10	11	12	1	2	3	4	5	6	7	8	9	10	11	12
A1					1		1				1		2					H0						
A2							1			1		2			1	H1	1							
B1													1							H3				
B2															1				1					
C1						H4		1																
C2																					1			
D1											1		1											
D2													1						1					
E1				1		1	1	H1						1								1		
E2								1		H1			1											
F1			1				2															H2		
F2																								

Table 3 Poisson cumulative distribution table

x	λ					
	1	2	3	4	5	6
1	0.74	0.41	0.20	0.09	0.04	0.02
2	0.92	0.68	0.42	0.24	0.12	0.06
3	0.98	0.86	0.65	0.43	0.27	0.15
4	1.00	0.95%	0.82	0.63	0.44	0.29
5	1.00	0.98	0.92	0.79	0.62	0.45
6	1.00	1.00	0.97	0.89	0.76	0.61

$$E = \frac{Removals}{365} * TAT. \quad (3)$$

From historic data, operator had 9 unscheduled removals per year, and longest TAT for the blade was 85 days. It can be calculated that demand equals two parts. Since the actual time at which a spare part is needed is stochastic, the Poisson conditional probability distribution is used to determine a realistic stock minimum for unscheduled removals. The demand calculated from expression 3 is a mean parameter for Poisson cumulative distribution table entry. Service Level (SL) is the probability of a part being available when needed. Reaching 100% SL would require a full duplication of all components in service. In practice, a target SL of 95% is used for No-Go items, or parts whose failure renders aircraft AOG until repair or replacement is performed (Macdonnell and Clegg 2006). SL is a probability parameter in Poisson cumulative distribution table. To satisfy 95% of demands for the spare blades for unscheduled replacement, operator needs to have four spare blades on stock at all times. The Poisson cumulative distribution table is partially shown in Table 3.

It can be deduced that with operator stock of 11 spare blades, the number of spare blades that can be used to accommodate scheduled replacements without affecting the 95% SL is 7. Since operator only has one spare hub, he will have to rely on other methods of procurement in case that unscheduled replacement of a hub is necessary at the same time a scheduled replacement is due, or reschedule the event.

4 Schedule for Staggered Replacement

To minimize maintenance costs, it is wise to group replacements of blades that are reaching the end of resource on the same position on the aircraft. After summing replacements due on each propeller positions shown in Table 2, we can identify total number of blades and hubs due for replacement each month. The data are show in Table 4.

Table 4 Total number of hubs and blades due for replacement each month

Month	Year 2020												Year 2021											
	1	2	3	4	5	6	7	8	9	10	11	12	1	2	3	4	5	6	7	8	9	10	11	12
Due	0	0	1	1	1	H5	5	H3	0	H2	2	2	6	1	2	H1	1	H0	2	H3	1	H3	0	0

Then, the staggering should be done with respect to spare hub availability. Since there is only one spare hub and TAT is two months, hub replacements need to be separated by two months to satisfy this limitation.

Next, a total number of the spare blades necessary at beginning of each month can be calculated. Knowing that the TAT for the blades is two months, and number of available spare blades on operator stock at the beginning of the schedule is six out of seven, it becomes clear at which points down the time line the number of spare blades would not be enough for scheduled replacements, or when the shortage will occur. Table 5 shows the number of blades needed each month and number of available spare blades including possible shortage situations denoted with minus sign in spare row cells. Before any scheduling is performed, we can count total shortage of 5 blades during two year interval.

At this point operator can chose whether he will exchange blades with part suppliers on the market or perform scheduling. To prevent additional costs, the operator can stagger the replacements. This will lead to some loss of resources which can be easily calculated and compared to exchange costs for the same situation.

Dynamic approach here refers to defining a problem similarly to the way as Dynamic Programing problems are defined, as described below. The problem of staggering the blades and hubs replacement has the following dynamic characteristics. Corresponding replacement timeline is divided into stages and at beginning of each stage the engineer has to make a decision. At each stage the needed number of blades for replacement is known. The spare blades availability determines maximum number of replacements at a given stage. Blades can be replaced at a given stage and before it but not after the due date. If the overhaul is moved up the schedule, staggering or earliness loss occurs. The goal of this schedule is to minimize losses created by early replacement, or to remove blades from service as late as possible. The problem described above is the problem of Backwards Scheduling.

Backward Scheduling refers to the process of scheduling the activities by commencing with the deadlines or latest possible finish date and time of activities and working backwards in reverse order of time, ultimately revealing the latest possible start date and time (Chugh 2009). Described scheduling problem can be solved by using of the Backwards Allocation algorithm. Algorithm steps can be defined as follows. Once the demand and availability of parts is known for each time slot, we can start to schedule the time slot. The algorithm starts with the last time slot and tries to allocate replacements as late as possible with respect to spares availability in the considered time frame. For example, if five replacements need to be done in April, the algorithm will try to allocate five spares. If less than five spare parts will be available in April, algorithm will push back replacements to the previous time slot. However, it will use all spares available in April before moving replacements backward. How far back replacements need to be moved to eliminate shortage depends on TAT. Solution for the problem identified in Table 5 as obtained by backward allocation algorithm is given in Table 6 where initial due dates and staggered plan are displayed alongside.

Table 5 Fleet replacement plan with calculated demand for spare blades at each interval, blades at overhaul (OH) and shortage situations denoted

Month	Year 2020												Year 2021											
	1	2	3	4	5	6	7	8	9	10	11	12	1	2	3	4	5	6	7	8	9	10	11	12
Due	0	0	1	1	1	H5	5	H3	0	H2	2	2	6	1	2	H1	1	H0	2	H3	1	H3	0	0
Spare	6	7	6	5	5	1	−3	−1	4	5	3	3	−1	0	4	4	5	6	5	2	3	3	4	7
OH	1	0	1	2	2	6	10	8	3	2	4	4	8	7	3	3	2	1	2	5	4	4	3	0

Table 6 Due dates for replacement and staggered replacement schedule as result of backward allocation algorithm

Year	2020								2021	
Month	5	6	7	8	9	10	11	12	1	2
Due	1	H5	5	H3	0	H2	2	2	6	1
Spare	5	1	−3	−1	4	5	3	3	−1	0
Staggered	4	H3	4	H3	0	H2	3	1	6	1
Spare	2	0	0	0	4	5	2	3	0	0

Year	2020			2021	
Month	10	11	12	1	2
Due	H2	3~~2~~	1~~2~~	6	1
Spare	5		3	0	0
Staggered	H2		1	6	1
Spare	5	TAT			0

Fig. 4 Application of backward allocation algorithm visualization

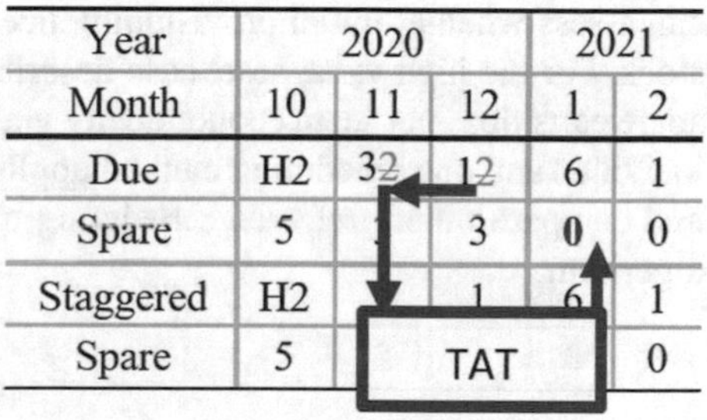

To accommodate 6 replacements in January 2021, one replacement had to be moved from December 2020 to November 2020. Backwards allocation algorithm steps are depicted graphically for January 2021 shortage (see Fig. 4).

To eliminate the shortage in middle of 2020, one replacement was moved from July to June, but then three replacements had to be moved from June to May. Staggered blades altogether will be removed 5 months before their due dates, leading to loss of 1083 FH resource and 1300 Money Units. Alternatively, operator could exchange 5 blades at the price of 2500 Money Units for each blade, leading to exchange costs of 12,500 Money Units.

Due to high complexity of the problem, the possibility of an engineer making the arbitrary plan that would generate minimal resource loss is highly unlikely, especially if he has to schedule the replacements on the larger fleet of aircraft. If used, the method can be a powerful tool for solving this backward scheduling problem. The operator can acquire the additional spare blades to increase the spare parts stock for 75,000 Money Units. If one compares this value to the calculated resource loss of 1300 Money Units for two-year period, it is obvious that cost of increasing the number of spares cannot be economically justified. Therefore, provisioning that would lead to increase in number of spare blades is not recommended for this case study.

5 Conclusion

This paper addresses the problem of scheduling preventative component maintenance activities. Upon examination of the problem dynamic properties were identified. The backward allocation algorithm has been applied to solve this problem. Due to fixed deadlines for component maintenance and costs incurred by performing early replacements, choosing backwards allocation for scheduling proved to be suitable method. For the application of the algorithm only initial plan of due dates is required and knowing the workshop TAT. A schedule can be easily made and resource loss for opportunity replacement calculated and minimized. With these data, the operator can assess whether initial provisioning needs to be revised and size the spare parts stock. For the high value asset case described in this paper, increasing the stock of spare parts does not seem economically viable. After problem constraints have been identified and one scheduling method applied, further research is needed to examine and compare other preferred scheduling algorithms with the backward allocation algorithm.

References

Ackert SP (2010) Basics of aircraft maintenance programs for financiers: evaluation & insights of commercial aircraft maintenance programs. https://aircraftmonitor.com/uploads/1/5/9/9/15993320/basics_of_aircraft_maintenance_programs_for_financiers___v1.pdf. Accessed 31 July 2020

Chugh T (2009) Backward scheduling—an effective way of scheduling warehouse activities. TATA Consult Serv Ltd White Pap

Dinis D, Barbosa-Póvoa A, Teixeira ÂP (2019) A supporting framework for maintenance capacity planning and scheduling: development and application in the aircraft MRO industry. Int J Prod Econ 218:1–15. https://doi.org/10.1016/j.ijpe.2019.04.029

Domitrović A, Bazijanac E, Čala I (2008) Optimal replacement policy of jet engine modules from the aircarrier's point of view. Promet Traffic Transp 20:1–9. https://doi.org/10.7307/ptt.v21i1.980

Erkoc M, Ertogral K (2016) Overhaul planning and exchange scheduling for maintenance services with rotable inventory and limited processing capacity. Comput Ind Eng 98:30–39. https://doi.org/10.1016/j.cie.2016.05.021

Hopp WJ, Kuo YL (1998) Heuristics for multicomponent joint replacement: applications to aircraft engine maintenance. Nav Res Logist 45:435–458. https://doi.org/10.1002/(SICI)1520-6750(199808)45:5%3c435::AID-NAV1%3e3.0.CO;2-4

Hosseini S, Kalam S, Barker K, Ramirez-Marquez JE (2019) Scheduling multi-component maintenance with a greedy heuristic local search algorithm. Soft Comput 7. https://doi.org/10.1007/s00500-019-03914-7

IATA (2015) Best practices for component maintenance cost management. https://www.iata.org/contentassets/bf8ca67c8bcd4358b3d004b0d6d0916f/cmcm-2nd-edition.pdf. Accessed 31 July 2020

IATA's Maintenance Cost Technical Group (2019) Airline maintenance cost: executive commentary edition 2019 (FY2018 data). In: MCTF Rep. https://www.iata.org/contentassets/bf8ca67c8bcd4358b3d004b0d6d0916f/mctg-fy2018-report-public.pdf. Accessed 31 July 2020

Macdonnell M, Clegg B (2006) Management of rotable aircraft spares inventory : development of a new solution [004-0250]. In: POMS conference. Production and Operations Management Society

Rutledge J (1997) Spare parts: cost benefit management. FAST Flight Airworth Support Technol 21:25–29

Sohn SY, Yoon KB (2010) Dynamic preventive maintenance scheduling of the modules of fighter aircraft based on random effects regression model. J Oper Res Soc 61:974–979. https://doi.org/10.1057/jors.2008.167

Monitoring Traffic Air Pollution Using Unmanned Aerial Systems

Marin Mustapić, Anita Domitrović, and Tomislav Radišić

Abstract Air Quality monitoring is traditionally performed by ground stations, and more recently by satellites. However, rapid, comprehensive data collection near pollution sources is not always feasible due to site complexity, relocation of pollution sources, or physical barriers. For these reasons unmanned aerial vehicles (UAVs) equipped with various air quality sensors are being introduced, offering new approaches and capabilities for monitoring air pollution, as well as studying atmospheric trends, such as climate change, while ensuring safety in urban and industrial areas. Due to small amount of data in the available literature, it is assumed that the research area is still at an early stage of development. Future research should include increase in flight endurance, payload, accuracy and sensitivity of sensors, and above all sensor calibration and flight planning in order to obtain the highest quality data.

Keywords Unmanned aerial systems · Unmanned aerial aircraft · Air pollution · Multirotors · Fixed wing aircraft · Air quality monitoring

1 Introduction

Polluted air has a detrimental effect on human health and negatively affects the environment. Over the past decades, emissions of many air pollutants in Europe have been significantly reduced, primarily by switching to more acceptable heating methods, which has led to improved air quality. However, the concentrations of air pollutants are still too high and air quality problems have not been eliminated.

A significant proportion of Europe's population lives in urban areas, where pollution levels of ozone, nitrogen dioxide and particulate matter are exceeded, which poses a serious threat to health and the environment. Air pollutants released at one

M. Mustapić (✉)
Croatia Civil Aviation Agency, Zagreb, Croatia
e-mail: marin.mustapic@ccaa.hr

A. Domitrović · T. Radišić
Faculty of Transport and Traffic Sciences, University of Zagreb, Vukelićeva 4, 10000 Zagreb, Croatia
e-mail: anita.domitrovic@fpz.unizg.hr

M. Petrović and L. Novačko (eds.), *Transformation of Transportation*, EcoProduction,
https://doi.org/10.1007/978-3-030-66464-0_11

location can reach other places through the atmosphere, where they can contribute to poor air quality. Therefore, air pollution is a problem both locally and globally (https://www.eea.europa.eu/hr/themes/air/intro).

Particulate matter, nitrogen dioxide and ground-level ozone are today considered to be the three pollutants that have the greatest impact on human health, especially in urban areas. Prolonged exposure to these pollutants leads to respiratory disorders and other related diseases. About 90% of the population of European cities is exposed to pollutants in concentrations above levels that are considered harmful to health (Report in the calculation of air pollutant emission in the territory of the Republic of Croatia 2019. (1990–2017)). A significant role in air pollution is played by the transport system with special emphasis on road traffic.

Air quality measurement is performed at stationary ground stations. Such conventional tracking systems of air pollutants have low spatial resolution. In order to increase the spatial resolution, the possibility of using unmanned aerial systems (UAS) was recognized.

With the advancement of technology in recent years, unmanned systems have gained popularity primarily due to the ability to provide more accurate information on pollution distribution in the vertical component of the atmospheric column, which is necessary to better understand the dispersion of pollution through atmospheric layers depending on meteorological parameters. Also, UAVs cover large areas and can collect samples at remote, dangerous, or hard-to-reach locations, increasing operational flexibility compared to conventional methods.

The aim of this paper is to present previous research, present the main advantages and disadvantages of the system and suggest further possibilities for improving the system.

2 Composition of Atmospheric Pollution and Effects on Human Health

2.1 The Origin of Pollution in the Atmosphere

The composition of the atmosphere is associated with emission processes that release large amounts of various compounds, gases and aerosols into the atmosphere. Also, the composition of the atmosphere at particular area is directly related to the earth's orography, the rotation of the earth, and to the macro and micro elements associated with the movement of air masses (Solomon and Sioutas 2008; Goldberg et al. 2001).

Sources of emissions can be natural, such as vegetation, the effects of deserts, volcanoes and open fires. In addition to natural, there are anthropogenic sources of pollution such as industrial plants, all forms of transport with special emphasis on road traffic and heating in households. Anthropogenic sources have a far more detrimental impact on health and the environment. Combustion processes release three main greenhouse gases into the atmosphere: carbon dioxide (CO_2), methane

(CH_4), and nitric oxide (N_2O). In addition, nitrogen monoxide (NO), sulphur dioxide (SO_2), and numerous other gases are formed. In addition to gases, particulate matter (PM) plays a very important role in air quality.

Suspended particles are all microscopic particles of matter in the range of 0.002–100 μm which, by the action of air currents, can float for a longer or shorter time until the final deposition on the ground, either by dry—gravitational or wet—precipitation deposition. Pollution of a certain area is related to meteorological conditions and the distribution and size of emissions on a local, regional and global scale.

It is very important to note that suspended particles are not only formed as a product of combustion, but also as a process of abrasion of vehicle tires, clutch system, vehicle braking system and abrasion of the pavement itself.

2.2 *Effects on Human Health*

2.2.1 Airborne Particles

According to the results of studies APHEA 1 and APHEA 2 (APHEA, Air Pollution and Health: A European Information System) in a large number of European cities, increasing concentration of PM10 airborne particles (particles with an aerodynamic diameter of less than 10 μm) by 10 $\mu g/m^3$ was associated with a daily increase in mortality of 0.6%, while mortality with increasing concentration by 100 $\mu g/m^3$ was as much as 6%.

There is a higher risk in cities with warmer climate and higher NO_2 concentrations. The risk is also higher due to the fact that PM10s of traffic origin are more toxic than those whose origin is not related to the combustion of motor fuels.

The health effects of long-term exposure to pollutants have been published in two studies conducted in the United States. Both studies document shorter life expectancies of people who lived in cities and were long-term exposed to average concentrations of suspended particles of two fractions: PM10 and PM2.5 (Dockery et al. 1993; Pope et al. 1995). The results of the study, prepared by the American Cancer Society, confirmed the results of previous research.

The study from the Netherlands in which 5000 adults who live near high-traffic roads was followed for 8 years showed a significant impact of PM on health (Hoek et al. 2002). Nowadays, Special attention is paid to the negative effects of ultrafine suspended particles, i.e. nanoparticles (diameter less than 0.1 μm) on health. Toxicological studies have shown that they are more toxic than other particles at the same mass doses received because the toxic effects are highly correlated with the number of particles and their surface area. Nanoparticles, by penetrating deep into the pulmonary interstitium can induce increased blood coagulation, thus increasing the risk of myocardial infarction (Peters et al. 1997).

3 Requirements for UAV Based Measurement Systems

Unmanned systems are much more versatile and simpler than other data collection systems such as conventional ground stations, manned aircraft systems and satellite measurement systems.

Atmospheric measurements at hard-to-reach locations are greatly facilitated by use of unmanned aerial vehicles (Barnhart 2012; Holland et al. 1992). Also, with the advancement of technology, the performance of the aircraft itself is increasing and the battery life is increasing, while the range of measuring instruments is expanding from year to year.

Unmanned systems are many times cheaper than conventional systems. As for the performance itself, unmanned systems can be divided into two main categories: fixed-wing aircraft and rotocopters. Some other concepts such as tied or free balloons can be found in the literature, however, they will not be the subject of this paper.

In the context of measuring air quality, it is very important to select an aircraft according to performance that is optimal for specific measurement or research objectives. Fixed-wing aircraft have number of advantages.

The basic characteristics of configuration of aircraft with fixed wings are simpler construction, and the possibility of achieving a longer flight at higher speeds and at higher altitudes, which allows collecting data from large areas in one flyby. Such aircraft are most often designed with high slender wings, which allows them to save energy by sailing. This achieves a better ratio of the aircraft's operating mass and payload, which allows to carry a heavier and more complex equipment. More efficient aerodynamics allows longer flight duration at higher speeds, which leads to collection of a large amount of data. Fixed-wing aircraft also have high wind resistance, making them ideal for such measurements.

Disadvantages of fixed-wing aircraft are the inability to collect data from one place and the inability to stay in place (hovering). Therefore, the obtained data have a significantly lower resolution. Likewise, fixed-wing aircraft often need an area (30–200 m) to obtain the required take-off speed or need to use special catapults to obtain the required speed, which greatly reduces the aircraft's flexibility in terms of transport and functionality (Wyllie 2001; Fahlstrom et al. 2012; Ozdemir et al. 2014).

Rotocopters such as multirotors (quadcopters, hexacopters, octocopters, etc.) generally have a lower maximum flight speed, but have significantly better performance when measuring air quality in smaller areas or in localized inspections. Also, their great advantage is the ability to hover and rise vertically, which gives a significantly better measurement resolution compared to aircraft with a fixed wing. Rotocopters are easier to operate, which can increase safety and reduce the risk of damage of the aircraft, payload or third parties. Rotocopters do not require special infrastructure for takeoff and landing, and depending on their size, they are easier to transport. It is very important that the position of the air quality sensor is optimally set to avoid the influence of rotors on the measurement quality.

Recently, more and more manufacturers are offering unmanned systems that can be used completely autonomously. Moreover, it is possible to combine multiple autonomous systems. This requires advanced algorithms, but the quantity and quality of data is greatly increased and at the same time the workload of the operator is reduced. In some countries, regulation allows one operator to monitor the operations of few unmanned systems simultaneously (Porat et al. 2016). Unmanned systems can be useful in measuring air quality parameters in critical situations or major disasters in industrial plants when it is extremely important to know the exact pollution parameters and the risk of entering in such contaminated areas is too high for any manned system.

3.1 Regulatory Requirements for Flying UAVs in Urban Areas

The ordinance on unmanned aircraft systems prescribes requirements for the safe use of Unmanned Aircraft with an operating mass of up to and including 150 kg, and requirements for the persons involved in the Unmanned Aircraft and Unmanned Aircraft Systems operations.

According to ordinance It is allowed to fly an Unmanned Aircraft in uncontrolled airspace up to 120 m above the surface or up to 50 m above the obstacle, whichever is greater and in controlled airspace outside a radius of 5 km from the aerodrome reference point up to 50 m above the surface. At a distance at least 3 km from thresholds and edges of an uncontrolled aerodrome runway, except where specific procedures for the flights of Unmanned Aircraft are depicted in the aerodrome's instructions for use.

Except vertical limitations there are also some limitations for horizontal distance from people, so it is possible to fly in such a way that the horizontal distance of an Unmanned Aircraft and assembly of people must be less than 50 m, except when an Unmanned Aircraft is taking part in a flying display. It is also allowed to fly in such a way that the horizontal distance from uninvolved people is not less than the flight altitude and not less than 5 m when the low-speed mode is activated on the Unmanned Aircraft, and when the maximum speed of 3 m/s is set, or 30 m of horizontal distance from people in all other cases (Ordinance on Unmanned Aircraft Systems "Official Gazette", number 104/18).

The ordinance on unmanned aircraft systems does not prescribes any requirements about autonomous unmanned aircraft flights but Croatian Civil Aviation Agency (hereinafter referred to as the Agency) has the right to request a special risk analysis from the aircraft operator and his proof that he can terminate the operation at any time and take control of the aircraft.

The remote pilot should not operate with more than one unmmaned aircraft at once. By way of ordinance, it is permissible to conduct flight operations with multiple unmanned aircraft at the same time, provided that the approval from the Agency has been obtained.

Since 2018, in Croatia UAV flights have been classified according to the weight of the UAV, maximum speed and flight location. The classification table is represented below (Table 1).

For safety reason airspace users need to be able to inform themselves about the airspace restrictions in real time. These issues are even more relevant in lower and uncontrolled airspace, as this tends to be more complex from the perspective of arrival and departure procedures for aerodromes.

To meet these conditions, Croatian Air Traffic Control made the software called AMC Portal. The AMC Portal is a web-based airspace management tool which provides relevant information to all airspace users in real time and enables direct communication between all airspace users and the ASM (Air Space Management) organization. By providing targeted information, it gives users the opportunity of making reservations of airspace by directly submitting a request and communicating with the ASM organization.

AMC Portal functionalities All airspace users can use the tool to review the current status of the airspace and the planned airspace activities/restrictions. While it also provides easy access to the relevant textual messages, it presents the practical implications for the airspace on different charts in 2D and 3D. This allows the users

Fig. 1 AMC Portal interactive interface

to quickly and effectively prepare themselves before conducting their own airspace activities. Using a built-in messaging system, all users can quickly respond to the ASM organization's tactical requests. Especially in case of contingency and a need for an urgent termination of activities, this is especially useful for UAS operators, and another way of putting safety first while taking into account user needs and requests. At planning level, it allows an efficient analysis of the user's requests and provides for an unambiguous way to issue the required approvals to conduct the planned activities in a safe manner (https://amc.crocontrol.hr/Portals/0/Slike_Vijesti/Dokumenti/4_%20AMC%20Portal%20Article.pdf) (Fig. 1).

4 Significant Research of Atmosphere and Air Pollution Using UAV

The development of technology, in order to apply much more sophisticated instruments that used network systems and direct data transmission, was described by Bates et al. (2013). The aim of the research was to generate a vertical profile of the atmosphere using an unmanned system. A total of 18 flights were performed. The payload was adapted to measure the concentration of particles and to measure the absorption and dispersion of light from aerosols (at three wavelengths) and to collect particles using 8 filters. The flight plan was to measure particle concentration while climbing to an altitude of 2700 m then descend to the altitude with the maximum measured concentration and perform sampling at that altitude. The aerosol concentration varied, with detection limits, from 0.04 to 0.51 $\mu g/m^3$. Using UAV, the scientists were able to measure the transport of particles and their distribution above the atmospheric boundary layer (ABL).

Harrison measured the horizontal, vertical, and temporal variability of suspended particles (PM) within the first 150 m of the atmosphere. The aim was to prove that UAVs could be an instrument for supplementing satellite data. The flights were performed in a cylindrical (spiral) shape, where each loop exceeded the previous one by 30 m, up to 150 m above sea level. The sensor used in the study was a particulate spectrometer with a suction probe placed under the cover, allowing for pure unobstructed collection of air samples. The mean concentration of PM 2.5 for three flights at different altitudes was 36.3 $\mu g/m^3$, and the highest concentration was below 10 m above sea level as seen in Fig. 2. The results showed a general vertical variation with a standard deviation of 3.6 $\mu g/m^3$ and proved that the concentration of PM 2.5 does not change significantly during the day (Harrison et al. 2015).

The influence of three-dimensional distribution of fine particulate matter (PM2.5) and meteorological elements was described by Si-Jia Lu et al. The objective of this study was to measure PM2.5 concentrations and meteorological data at 300–1000 m altitude using an UAV equipped with mobile instruments. The study was conducted in a 4×4 km^2 space in Lin'an, Yangtze River Delta (YRD), China. The UAV was operated repeatedly for four times in one day along the designed route

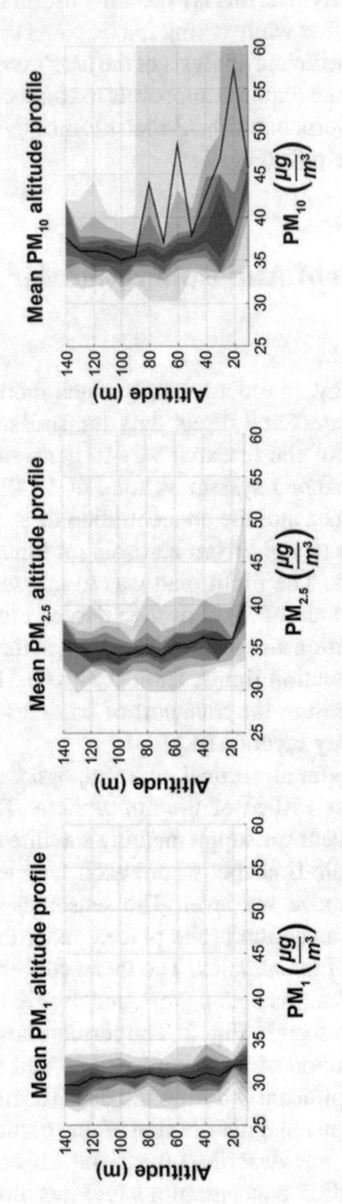

Fig. 2 Correlation between mean (black line) and total number of particles (red shaded) and altitude (Harrison et al. 2015)

Table 1 Flight operations categorization, Ordinance on Unmanned Aircraft Systems Pravilnik o sustavu bespilotnih letjelica. (NN 104/18)

Flight operations category	Unmanned aircraft		Conduct of flight operations		Requirements for the remote pilot		Requirements for the operator	
	Unmanned aircraft operating mass	The highest speed of an unmanned aircraft according to the manufacturer's technical specifications	Part of the day	Area of flight operations	Minimum age	Necessity of taking theoretical/Practical exams	Obligation to register/Approve the operator	Operator documentation
A	**OM < 250 g**	<19 m/s	Day and/or night	Unpopulated and/or populated area	Not applicable	Not applicable	Not applicable	Not applicable
B1	**250 g < QM < 900 g**	<19 m/s	Day	Unpopulated area	14 years, or less than 14 years but under adult supervision	Not applicable	Not applicable	Not applicable
B2	**OM < 5 kg**	Not applicable	Day and/or night	Unpopulated and/or populated area	16 years	Not applicable	Registration	Not applicable
C1	**5 kg < OM < 25 kg**	Not applicable	Day	Unpopulated area	18 years	Passed theoretical knowledge examinations provided by Agency	Registration	Not applicable

(continued)

Table 1 (continued)

Flight operations category	Unmanned aircraft		Conduct of flight operations		Requirements for the remote pilot		Requirements for the operator	
	Unmanned aircraft operating mass	The highest speed of an unmanned aircraft according to the manufacturer's technical specifications	Part of the day	Area of flight operations	Minimum age	Necessity of taking theoretical/Practical exams	Obligation to register/Approve the operator	Operator documentation
C2	**5 kg< OM < 150 kg**	Not applicable	Day and/or night	Unpopulated and/or populated area	18 years	(a) Passed theoretical knowledge examinations provided by Agency (b) Demonstration of flight preparation and flying	Approval	(a) Operations Manual (b) Flight records (c) Safety risk assessment

spirally from the ground to 1000 m altitude with a total of 8 layers and a 100 m interval between 15 two adjacent layers for five days from 21st August 2014 to 2nd February 2015. PM2.5, air temperature, relative humidity, dew point temperature and air pressure were measured during the data collection. The study results indicated that the PM2.5 concentrations decreased with altitude at 300–1000 m and the variations of PM2.5 with altitude in morning flights were much bigger than in afternoon flights. Besides, the PM2.5 concentration levels in morning flights were generally lower than in afternoon flights. PM2.5 concentrations were positively correlated with dew point temperature and pressure, but positively correlated with relative humidity only on pollution days in autumn or winter. The vertical gradient of PM2.5 concentrations was small in pollution days compared with on clean days. These findings provide the key theoretical foundation for PM2.5 pollution forecast and environmental management (Lu et al. 2016) (Fig 3).

The potential for collecting air quality data with high spatial and temporal resolutions was described by Qijun Gu et al. The aim of study was to design and develop a modular UAV-based platform capable of real-time monitoring of multiple air pollutants. The system comprised five modules: the UAV, the ground station, the sensors, the data acquisition (DA) module, and the data fusion (DF) module. The hardware was constructed with off-the-shelf consumer parts and the open source software Ardupilot was used for flight control and data fusion. The prototype UAV system was tested in representative settings. Results showed that UAV platform can fly on pre-determined pathways with adequate flight time for various data collection missions. The system simultaneously collected air quality and high precision X-Y-Z data and integrated and visualizes them in a real-time manner. While the system could accommodate multiple gas sensors, UAV operations may electronically interfere with the performance of chemical-resistant sensors. Prototype and experiments prove the feasibility of the system and show that it features a stable and high precision spatial-temporal platform for air sample collection (Gu et al. 2018).

Further research went mostly in the direction of measuring air quality parameters in industrial zones and urban areas, where majority of the world's population lives, which is directly exposed to pollution. Therefore, there is a need to measure atmospheric parameters near major roads, in order to get more accurate picture of influence of the transport system in the total pollution of boundary layer.

5 Current Situation and Possibilities for Further Research

The use of unmanned systems in air quality research is widely used: from measuring the amount and composition of gases at different altitudes, comparing the obtained data with conventional stations on earth, to fully autonomous monitoring of air pollution. The accelerated development of technology greatly contributes to all this, primarily in terms of the physical reduction of the size of instruments for measuring air quality, which increases the possibilities of using the aircraft themselves (Koppmann 2008).

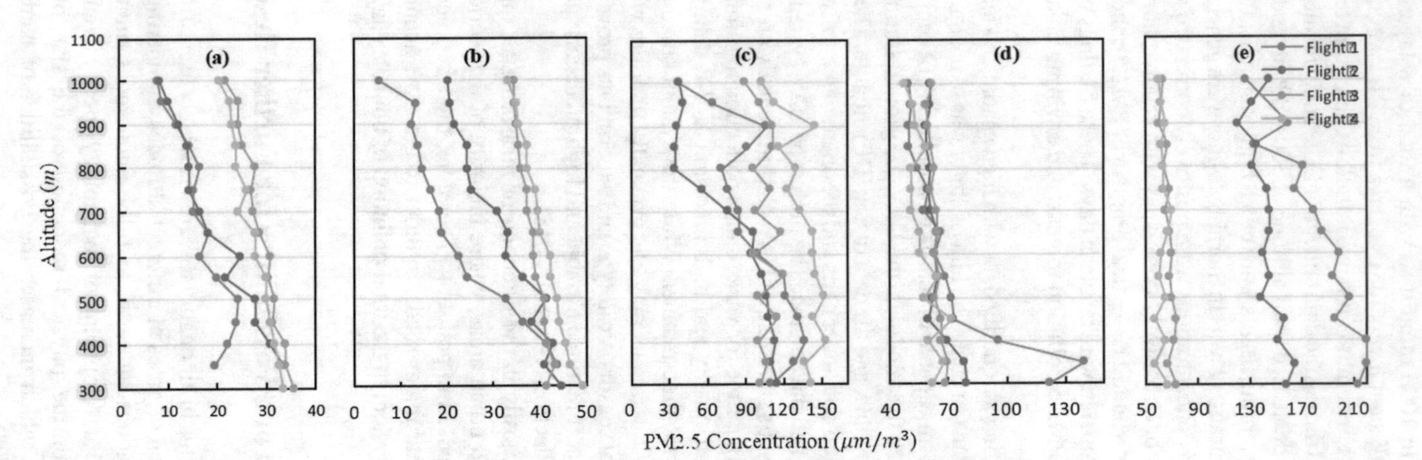

Fig. 3 PM 2.5 mass concentration profiles of all flights. Flights 1-4 during a monitoring day are marked with blue, red, gray and yellow, respectively. **a** 2014/8/20; **b** 2014/10/11; **C** 2014/11/14; **d** 2014/12/12; **e** 2015/2/5. (Lu et al. 2016)

Ultra-light instruments, such as laser-based dust sensors and gas sensors with low energy consumption requirements are evolving at a high rate, primarily due to the need to investigate volcanic emissions (Andres et al. 2010). At the same time, navigation systems are being developed that enable autonomous locating of pollution (Neumann et al. 2016; Porter and Vasquez 2006). Unmanned systems can also supplement the data of existing earth stations to improve the accuracy of data that directly enters in the air quality modelling process.

A review of the available literature shows that the use of unmanned systems in the context of measuring air quality is justified and that there is the possibility of further research. The possibilities of future research can be divided into three categories:

1. Direct impact of meteorological parameters on air pollution profile in upper air
2. Sampling air on higher altitude to calculate the amount of anthropogenic pollutants, and
3. The impact of aircraft turbulences on data quality.

Meteorological parameters such as air temperature, relative humidity, air pressure, and wind speed and direction are known to play an important role in the study of air pollution. The meteorological parameters determine how long a certain particle will remain in the atmosphere, and to which vertical and horizontal limits it will move. Previous research on the dependence of traffic pollution on meteorological conditions shows that the influence of meteorological conditions on concentration of total air pollution is much greater than the impact of traffic. Research is mostly made with instruments on the ground and do not have a vertical component which allows unmanned systems to continue research (Bencetić Klaić et al. 2009).

When it comes to the total amount of particles in the atmosphere, then it is important to know whether these are naturally occurring particles such as plant aeroallergens, desert dust, sea salt particles, etc. or anthropogenic particles caused by the human activities. There is a clear justification for the use of unmanned systems because they have ability to sample the air in almost all places, which is not the case with conventional devices.

In order to obtain quality data, it is necessary to investigate where and in what way optimally position the measuring instrument in order to avoid the influence of the propeller on the accuracy of data. Also, it is necessary to plan the flight route so that data are not collected from the atmosphere whose condition was previously disturbed by the passage of the aircraft itself.

Most of previous researches did not take into account the positioning of the measuring instrument as well as the choice of flight route, and in this segment could be room for further research.

Future air quality instruments should also be equipped with real-time self-calibration systems, while improving the autopilot system would greatly help in flight planning accuracy, thus optimizing the sampling process (Ippolito et al. 2009). Currently, the software support and quality of aircraft control in rotocopters is significantly more developed than the system for fixed wing aircraft. The reason for this stems from the widespread use of rotocopters in recent years, primarily for aerial photography purposes. For that reason, rotocopters are used in monitoring and measuring air quality wherever their lower persistence meets measurement needs.

6 Conclusion

Unmanned aerial vehicles can offer extremely high resolution of spatial sampling, which is not feasible with any other method used so far. Some systems are more flexible than others, due to their ability to carry heavier loads or greater flight endurance, however, the future of their use is almost guaranteed.

At the same time, advances in technology in fields such as chemistry, physics and information technology result in smaller and lighter measuring instruments, greater sensitivity and remote working capabilities.

Key limiting factors in the development of measuring sensors include: power, mass and size, because they are closely related to the type of platform (fixed wing or rotocopter), the choice of propulsion group (electric or internal combustion drive) and the type of operation.

In the last few years, the number of papers about use of unmanned systems has significantly increased, and the scientific community has recognized the possibilities of applying the system. However, there are still no concrete or agreed procedures for the operational application of unmanned systems. It is evident from the available literature that there is no uniform procedure for sampling the atmosphere.

The first problem occurs when setting up and calibrating the sensor, where several authors approach the measurements in a different way, which is visible in the obtained results i.e. it can be said that the results oscillate too much. In addition to calibration and sensor placement, there is a problem with flight route selection. To obtain quality results, it is extremely important to repeat the procedure in an identical way, which is often not the case. In the direction of creating a basis for the unification of measurements, numerous possibilities for further research have been opened.

References

Andres Diaz J, Pieri D, Arkin CR, Gore E, Griffin TP, Fladeland M, Bland G, Soto C, Madrigal Y, Castillo D et al. (2010) Utilization of in situ airborne ms-based instrumentation for the study of gaseous emissions at active volcanoes. Int J Mass Spectrom 295:105–112. Available from: https://link.springer.com/article/10.1007/s13361-014-1058-x

Barnhart RK (2012) Introduction to unmanned aircraft systems. CRC Press, Boca Raton, FL, USA. Available from: http://rahauav.com/Library/Unmanned%20Vehicles/Introduction_to_Unmanned_Aircraft_Systems.pdf

Bates TS, Quinn PK, Johnson JE, Corless A, Brechtel FJ, Stalin SE, Meinig C, Burkhart JF (2013) Measurements of atmospheric aerosol vertical distributions above svalbard, norway, using unmanned aerial systems (UAS). Atmos Meas Tech 6:2115–2120. Available from: https://www.atmos-meas-tech.net/6/2115/2013/

Bencetić Klaić Z, Ružman K, Smiljanić I, Stojnić M (2009) Utjecaj gustoće prometa i meteoroloških čimbenika na koncentraciju PM1.0 čestica u zraku, Zagreb. Available from: https://apps.unizg.hr/rektorova-nagrada/javno/stari-radovi/230/preuzmi

Dockery DW, Pope CA III, Xu X, Spengler JD, Ware JH, Fay ME, Ferris BG, Speizre FE (1993) An association between air pollution and mortality in six US cities. N Engl J Medicine 329:17531759. Available from: https://pubmed.ncbi.nlm.nih.gov/8179653/

Fahlstrom PG, Gleason TJ (2012) Launch systems. In: Introduction to UAV systems, 4th ed, Wiley, Hoboken, NJ, USA, pp 249–260. Available from: shorturl.at/esPSZ

Goldberg MS, Burnett RT, Bailar JC, Brook J, Bonvalot Y, Tamblyn R, Singh R, Valois MF, Vincent R (2001) The association between daily mortality and ambient air particle pollution in Montreal, Quebec 2.Cause-specific mortality. Environ Res 86, 26–36. Available from: https://doi.org/10.1006/enrs.2001.4243

Gu Q, Michanowicz DR, Jia C (2018) Developing a modular unmanned aerial vehicle (UAV) platform for air pollution profiling. Sensors 18(12):4363. Available at: https://doi.org/10.3390/s18124363

Harrison WA, Lary DJ, Nathan BJ, Moore AG (2015) Using remote control aerial vehicles to study variability of airborne particulates. Air Soil Water Res 8:43–51. Available from: https://utd-ir.tdl.org/bitstream/handle/10735.1/5343/NSM-3122-5274.55.pdf?sequence=1&isAllowed=y

Hoek G, Brunekreef B, Goldbohm S, Fischer P, Brandt PA (2002) Association between mortality and indicators of trafficrelated air pollution in the Netherlands: a cohort study. The Lancet 360:12031209. Available from: https://cris.maastrichtuniversity.nl/en/publications/association-between-mortality-and-indicators-of-traffic-related-a

Holland G, McGeer T, Youngren H (1992) Autonomous aerosondes for economical atmospheric soundings anywhere on the globe. Bull Am Meteorol Soc 73:1987–1998. Available from: shorturl.at/osQT0

https://www.eea.europa.eu/hr/themes/air/intro (June 2018.)

Ippolito C, Fladeland M, Yoo Hsiu Y (2009) Applications of payload directed flight. In: Proceedings of the 2009 IEEE aerospace conference. Big Sky, MT, USA, 7–14 March 2009, p 1

Izvješće o proračunu emisija onečišćujućih tvari u zrak na području republike hrvatske 2019. (1990–2017). Available from: http://www.haop.hr/sites/default/files/uploads/dokumenti/011_zrak/Izvjesca/Izvjesce_o_proracunu_emisija_oneciscujucih_tvari_u_zrak_na_podrucju%20RH_2019.pdf

Koppmann R (2008) Volatile organic compounds in the atmosphere. Wiley, Hoboken, NJ, USA. Available from: https://www.wiley.com/en-us/Volatile+Organic+Compounds+in+the+Atmosphere-p-9781405131155

Lu S-J, Wang D, Li X-B, Wang Z, Gao Y, Peng, ZR (2016) Three-dimensional distribution of fine particulate matter concentrations and synchronous meteorological data measured by an unmanned aerial vehicle (UAV) in Yangtze River Delta, China, Atmos Meas Tech Discuss. Available from: https://doi.org/10.5194/amt-2016-57

Neumann PP, Bennetts VH, Lilienthal AJ, Bartholmai M (2016) From insects to micro air vehicles—A 12 comparison of reactive plume tracking strategies. In: Intelligent autonomous systems 13; Springer, Berlin, pp 1533–1548. Available from: https://www.semanticscholar.org/paper/From-Insects-to-Micro-Air-Vehicles-A-Comparison-of-Neumann-Bennetts/aa614a08a027fb2adb75c1b4b5ae1d9e58d8bce4

Ozdemir U, Aktas Y, Vuruskan A, Dereli Y, Tarhan A, Demirbag K, Erdem A, Kalaycioglu G, Ozkol I, Inalhan G (2014) Design of a commercial hybrid vtol UAV system. J Intell Robot Syst 74:371–393. Available from: https://link.springer.com/article/10.1007/s10846-013-9900-0

Peters A, Wichmann E, Tuch T, Heinrich J, Heyder J (1997) Respiratory effects are associated with the number of ultrafine particles. Am J Respir Crit Care Med 155:13761383. Available from: https://pubmed.ncbi.nlm.nih.gov/9105082/

Pope CA III, Thun MJ, Namboodiri MM, Dockery DW, Evans JS, Speizer FE, Heath CJ (1995) Particulate air pollution as a predictor of mortality in a prospective study of U.S. adults. Am J Resp Crit Care Med 151:669674. Available from: http://www.scientificintegrityinstitute.org/Pope1995.pdf

Porat T, Oron-Gilad T, Rottem-Hovev M, Silbiger J (2016) Supervising and controlling unmanned systems: a multi-phase study with subject matter experts. Front Psychol 7. https://doi.org/10.3389/fpsyg.2016.00568

Porter MJ III, Vasquez JR (2006) Bio-inspired, odor-based navigation—Art. No. 62280v. In: Schum K, Sisti AF (eds) Modeling and simulation for military applications, vol 6228. SAGE Publishing, London, UK, pp V2280–V2280

Solomon PA, Sioutas C (2008) Continuous and semicontinuous monitoring techniques for particulate matter cmass and chemical components: a synthesis of findings from epa's particulate matter supersites program and related studies. J Air Waste Manag Assoc 58:164–195. Available from: https://www.researchgate.net/publication/5534575_Continuous_and_Semicontinuous_Monitoring_Techniques_for_Particulate_Matter_Mass_and_Chemical_Components_A_Synthesis_of_Findings_from_EPA's_Particulate_Matter_Supersites_Program_and_Related_Studies

Wyllie T (2001) Parachute recovery for UAV systems. Aircr Eng Aerosp Technol 73:542–551. Available from: https://www.emerald.com/insight/content/doi/10.1108/00022660110696696/full/html?skipTracking=true

Instruments for Career Development in the Air Transport Industry

Sorin Eugen Zaharia, Adina Petruta Pavel, Casandra Venera Pietreanu, and Steliana Toma

Abstract The complex and continuously increasing air transport sector is very dynamic, being directly influenced by societal changes such as demographic aspects, globalization, urbanisation, digitalisation, climate change and even the current Covid-19 global crisis. Moreover, the new digital technologies applied in aviation require the evolution of current jobs competences and consequently the skills of graduates in order to practice the new occupations. Some transversal skills and soft skills allowing to evolve in this new digital world become crucial. After studying ICAO and EC documents on qualifications and focus groups with stakeholders, the first conclusion is that we are confronted with a strong growth in recruitment needs, in many cases for new occupations and a lack of competent profiles, especially for interdisciplinary ones. For supporting the employment in air transport, the carrier development and professional orientation, we propose some instruments: 2 methodologies development within a European project and on-line platforms offering information on jobs and qualifications in aviation. We will highlight the role of the developed methodologies, that of regulation, to generate minimum criteria, standards and recognition procedures and, thus, to be used as support by all stakeholders. For people to know what skills will be necessary for future occupations, they need to know where to get it from and how to use it. Here comes the communication as a key element; one of the solutions to find the jobs best suited to professional expectations and at the same time to reduce the gap between employers' expectations regarding the skills and knowledge or skills of graduates is to use professional orientation platforms as faithful to

S. E. Zaharia · C. V. Pietreanu (✉)
Faculty of Aerospace Engineering, University Politehnica of Bucharest, Splaiul Independenţei 313, 060042 Bucharest, Romania
e-mail: casandra.pietreanu@yahoo.com

S. E. Zaharia
e-mail: sorin.zaharia@gmail.com

A. P. Pavel · S. Toma
University Politehnica of Bucharest, UNESCO Chair "Engineering for Society", Splaiul Independenţei 313, 060042 Bucharest, Romania
e-mail: adinapavel@gmail.com

S. Toma
e-mail: steliana@tif.ro

M. Petrović and L. Novačko (eds.), *Transformation of Transportation*, EcoProduction,
https://doi.org/10.1007/978-3-030-66464-0_12

the professional challenges of twenty-first century. Long-term communication, guidance and connection between education, training providers and employers will be sustained through a European network for aviation education and training (ENATE).

Keywords Air transport · Career development · Recognition · Sectoral qualification framework · ENATE

1 Current Economic and Social Context for Air Transport Sector

The air transport industry has an important contribution to global economic and social development, currently supporting a total of 65.5 million jobs globally, of which 10.2 million direct jobs and generating a total economic contribution of USD 2.7 trillion to the gross domestic product (GDP). Airlines, air navigation service providers and airports directly employ around 3.5 million people and the civil aerospace contains 1.2 million employees. Another 55.3 million jobs supported by air transport sector are indirect, induced and tourism-related (Fig. 1). This trend will continue, both air passenger traffic and air freight traffic are expected to more than double in the next two decades. Forecasts indicate that in 2036, aviation will provide 98 million jobs and generate USD 5.7 trillion in GDP (ICAO 2019).

Despite these statistics, since the beginning of 2020, several countries worldwide were affected by COVID-19 and even shut down borders and limited domestic travel. Thus, cancelling almost all flights to control the spread of the virus has affected the entire airline industry globally. Flights were 70% lower at the start of Q2 and further decline is possible as restrictions continues in several regions. Due to the coronavirus outbreak, airport passenger traffic in Europe drastically decreased by 90% compared to the same day in 2019 (ACI 2020a) (Fig. 2).

According to world data from Airports Council International (2020), global passenger traffic declined by an unprecedented −94.4% year-over-year in April, followed an already dramatic drop of −55.9% in March, representing the worst decline of global passenger numbers in the history of the aviation industry (ACI 2020a, b).

Aviation sector	Employment (jobs)	Economic benefit (GDP)
Aviation direct	10.2 million	$704.4 billion
Aviation indirect	10.8 million	$637.8 billion
Aviation induced	7.8 million	$454.0 billion
Tourism catalytic	36.7 million	$896.9 billion
Total	**65.5 million**	**$2.7 trillion**

Fig. 1 Employment and economic benefits generated by aviation sector. Reproduced from ICAO (2019)

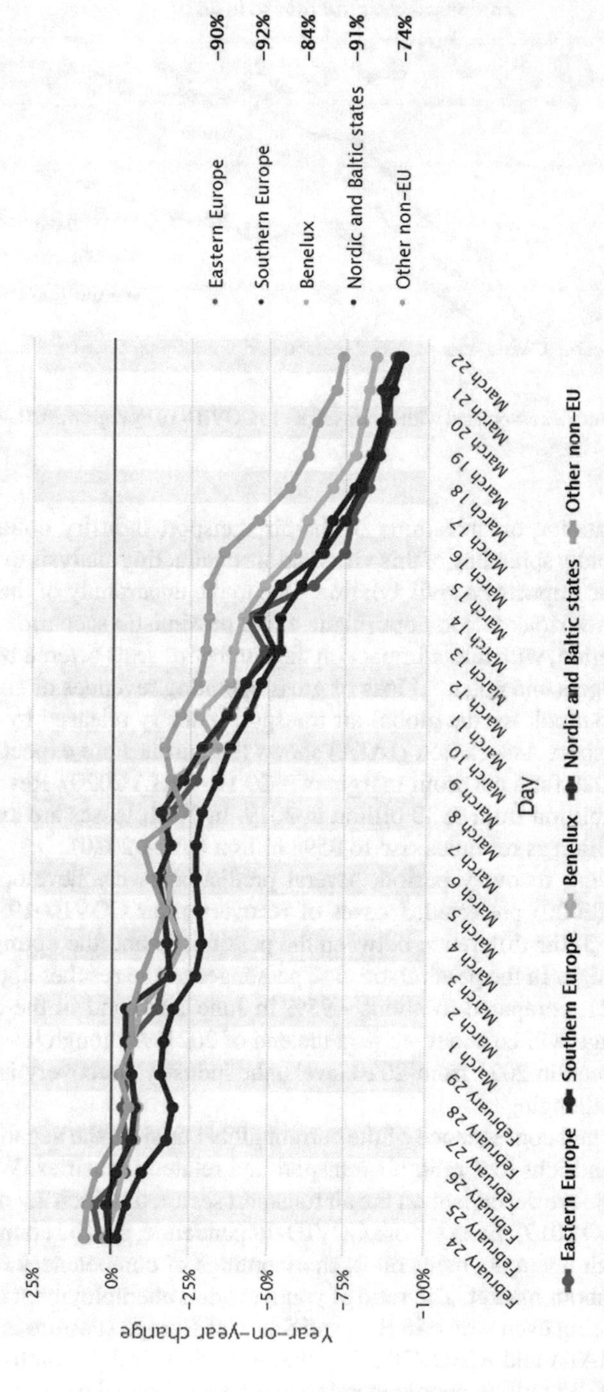

Fig. 2 ACI-Impact of COVID-19 on European airport passenger traffic (ACI 2020a and Statistica Research Department 2020)

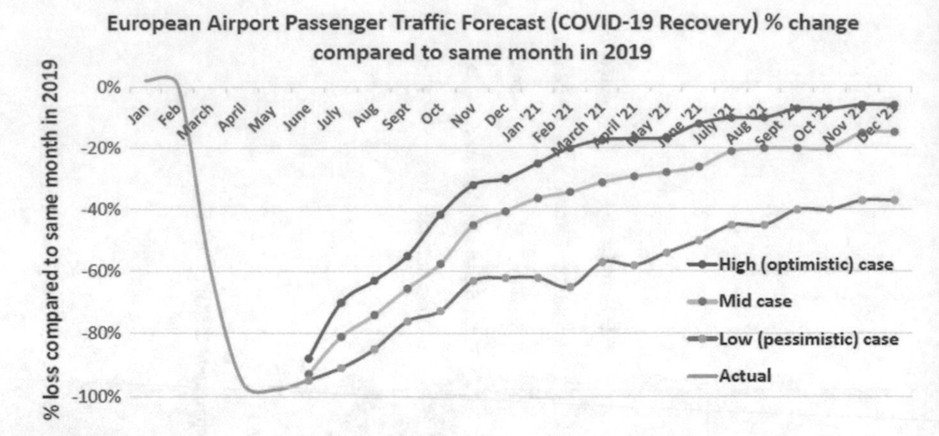

Fig. 3 Job loss in industries associated with air travel due to COVID-19 by region 2020. Reproduced from ACI (2020c)

Several international organizations in the air transport industry collaborate in monitoring the rapidly spreading of this virus and are conducting analysis to assess the potential economic impacts on civil aviation. Due to the uncertainty of the situation and depending on various factors, optimistic and a pessimistic scenarios are being taken in consideration, with direct impact on the number of seats offered by airlines, number of passengers and potential loss of gross operating revenues of airlines.

The financial outlook for the global air transport industry released by the International Air Transport Association (IATA) shows that airlines are expected to lose $84.3 billion in 2020 for a net profit margin of −20.1% (IATA 2020). Revenues will fall 50% to $419 billion from $838 billion in 2019. In 2021, losses are expected to be cut to $15.8 billion as revenues rise to $598 billion (IATA 2020).

For the following recovery period, several predictions were developed. IATA (2020) and ACI (2020) predicted 3 cases of recovery after COVID-19 outbreak, illustrated in Fig. 3, the difference between the pessimistic and the optimistic case being extremely high. In the pessimistic case passenger traffic reaches about −37% at the end of 2021, compared to about −95% in June 2020 and in the optimistic case, the percentage will be about −5% at the end of 2021. Although losses will be significantly reduced in 2021 from 2020 levels, the industry's recovery is expected to be long and challenging.

Another important consequence of the current global crisis is the negative impact on employment and jobs loss in in air transport and related industries. Worldwide, 65.5 million people are dependent on the air transport sector, of which 2.7 million are airlines jobs (ICAO 2019). Even before COVID-19 pandemic, the EU countries were facing rising youth unemployment rates, shortcomings of competences relevant to the air transport labour market. The ratio of youth to adult unemployment is between 2 and 3, figures being even worse in East and South of Europe (Pastore 2018).

According to IATA and ICAO (2020), if the restrictions last 3 months, it is estimated that approx. 35 million people working in air travel related industries can lose

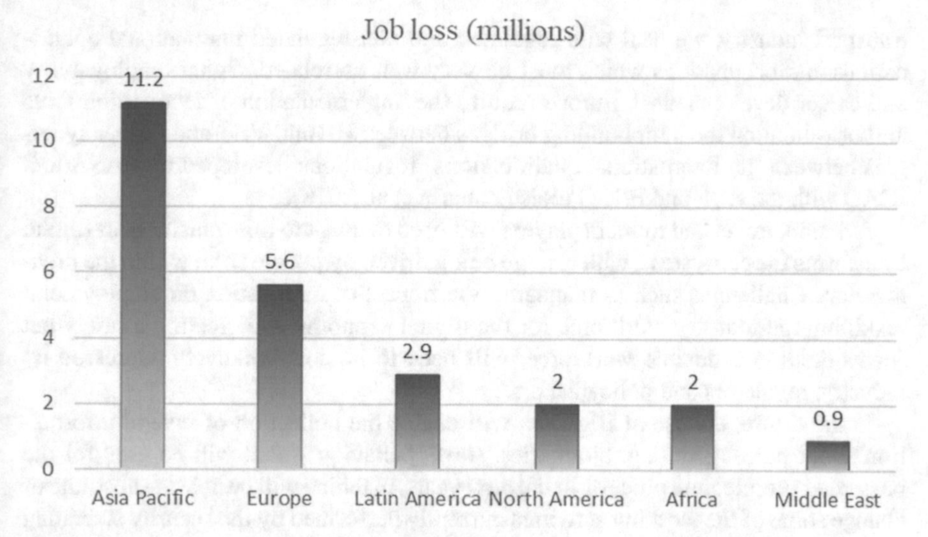

Fig. 4 Job loss in industries associated with air travel due to COVID-19 by region 2020. Reproduced from Pearce (2020)

their jobs due to the coronavirus outbreak and 5.6 million people in Europe (Fig. 4), the tourism and travel industry being among those the hardest affected. In the same scenario, airlines are expected to see full year passenger revenues fall by $252 billion (−44%) in 2020 compared to 2019 (ICAO 2020). The second quarter is considered the most critical with demand falling 70% and airlines burning through $61 billion in cash (Pearce 2020).

Considering the above-mentioned facts, but also considering the major societal changes, all the processes and activities related with air transport industry will need to adapt.

Many of the major societal changes currently affecting the transportation system are expected to have a significant impact on jobs. The fast-changing job market will require new sets of skills mostly related to problem solving, critical thinking and creativity. To face these challenges, industries and educational bodies need to collaborate to make the school-to-work transition as smooth as possible and to be able to prepare future generations for the world of work.

2 The Need for New Qualifications and Jobs Related in Air Transport Industry

Understanding future needs and necessary skills for air transport labour market is essential for shaping education and training policies in this sector. In the current society, digital transformation, globalization and job mobility demands a different set of skills from workers, this leading to the creation of new types of jobs. In air transport industry, we deal with regulated and non-regulated international qualifications and occupations which must be very well correlated. Better employability and career development in Europe require the implementation of recognition tools and/or validation tools for building bridges between HE and VET and further synergies between the international qualifications' regulations developed by EASA and ICAO with the EQF and ESCO tools (Zaharia et al. 2018).

Aviation, travel and tourism players will need to adapt to this transition, as digital transforms the ecosystem, with change being driven by people from within the organization. Challenges such as managing the impact of automation on employment, reskilling the industry workforce for the digital economy and creating a safety net for workers in a flexible workforce, will need to be collaboratively addressed by industry, regulators and policymakers.

In the future, the use of Big Data will enable the collection of several information about passengers (e.g. biometrics, travel behaviour), that will be used for the passenger preselection process. In the next years, machine will be able to substitute or change some of the working activities currently performed by the security screening officers at airports.

Within the European Erasmus + Project "Knowledge Alliance in Air Transport (KAAT)", an extensive analysis of current occupations was carried out to provide insights into the knowledge, skills and competences required by the current and emerging employers of the aviation sector, and to better align educational programmes to the needs of future generations of workers. A comprehensive review of the European labour market in the aviation sector was carried out. The main output of this analysis was the development of a sectorial breakdown, within which around 120 occupations were identified and described by providing insights into the key competences, tasks, responsibilities and learning outcomes required by the current employers (KAAT 2018). Considering the results of this research, among the identified emerging occupations are: Software and AI engineers, Security and cyber security experts, Remote tower controllers, Big Data and analytics experts, Unmanned traffic controllers, AI engineers/VR experts, Robotics engineers, Legal services personnel and ethics and privacy protection specialists.

These new occupations will require new additional skills and not to replace the existing ones. This means adding new interdisciplinary qualifications and many learning returns in order to upgrade the skills and knowledge.

These new occupations ask for new competencies. The aviation engineer of the future will require a mixture of technical and soft skills that are related to the current context of digitalisation and increasingly rapid technological change, including: digital competencies (advanced analytics and big data, cloud and as a service platform, mobility, etc.); design thinking; entrepreneurial thinking; cyber security skills; skills related to virtual/augmented reality.

The concerns about environmental issues and "greenization" of the airports will also generate new jobs and consequently lead to new qualifications. In this regard, some new identified jobs are: energy and maintenance engineer, climate change reversal specialist, consumer energy analysts, electrical engineer/alternative vehicle developers, battery technician or solar flight specialists. Many people involved in the field of the environment have to follow new qualifications or a technical training within the lifelong learning. Also, the core curriculum of some existing qualifications has to be improved following the new competences asked by new jobs or by the change of existing ones.

3 Instruments for Regulation, Guidance and Career Development in Air Transport

This paper present some of the key results of the research conducted on occupations and qualifications in air transport under the Erasmus + project "Knowledge Alliance in Air Transport" (KAAT) coordinated University POLITEHNICA of Bucharest and involving 5 partners 11 associates from 5 countries.

For supporting the employment in air transport industry, the carrier development and professional orientation of candidates, we proposed the following instruments: as regulation instruments, one methodology for sectoral qualifications framework and one methodology for recognition of prior learning and work experience in aviation, both developed in KAAT project and as guidance and communication instrument, examples of relevant on-line platforms offering information on jobs and qualifications in aviation.

Recognition of competences and correlation of learning outcomes may provide support for designing coherent professional pathways and complementarity of general training in the aviation field provided by high schools or universities and specific training for occupations in the sector.

When we refer to the air transportation, this subject does not follow the usual educational pattern due to the strict requirements of the industry which is imposing the necessity of being licensed and certified in order to be able to work in such an environment. The major question arising is to what extent are universities capable of providing competent graduates ready for direct insertion in this highly regulated field of work? Let see what the main pathways for education and training are.

There are two main pathways: the academic one, which consists in bachelor, master and doctorate study programs, which can be followed by or can alternate with postgraduate trainings; the vocational one, ensured by various training providers, such as airlines, handling companies, regulatory bodies, private trainers, in compliance with European Aviation Safety Agency (EASA) and International

Civil Aviation Organization (ICAO) regulations (KAAT 2019). Our concern is to create as many bridges as possible between the two pathways.

3.1 Methodology for Sectoral Qualifications Framework in Air Transport

Following the researches and reports prepared by KAAT consortium, a Methodology for Sectoral Qualifications Framework in Aviation was developed, aiming to help the universities and training providers to describe their qualifications in terms of learning outcomes; as well, it helps all stakeholders from aviation to understand or to create the links between academic and vocational pathways, and the links between regulated and non-regulated qualifications.

At the same time, the Methodology helps employees to understand the connections between different qualifications and to know how could acquire an international recognition of their study and qualifications. Also, it offers the tools for recognition, the links between different levels of qualifications and the correlations between different countries to ensure a European recognition.

Sectoral qualification framework for air transport (SQFAT), developed within KAAT Project, adopted the principles of the new learning paradigm, shifting the focus from inputs to outputs and on learning outcomes. At the same time, SQFAT proposes a dynamic and specific relationship between learning outcomes and professional competences.

The SQFAT methodological model is a reference framework developed for the analysis, description and interpretation of qualifications in the air transport sector. This model is compatible with the European Qualifications Framework (EQF), especially with the learning outcomes specified by the EQF for qualification levels 5, 6, 7 and 8. The structure and contents of the model harness on the descriptors of the EQF, as well as on the content of some models that have been already appreciated by European experts.

The essential components of SQFAT methodological model are a SQFAT Matrix (Fig. 5) and 2 other tools, a descriptive one and one for presenting the relation between competences, learning outcomes and study programs' curricula. There are 2 main categories of competences, professional competences and transversal competences. Professional competences are specific to a professional activity and contain the following learning outcomes: knowledge, skills, autonomy and responsibility, social and personal development. Transversal competences refer to social and personal development and are expressed in terms of the following descriptors: social interaction and life and career management. The qualifications' matrix refers to these competencies, includes qualification levels, learning outcomes, generic descriptors associated to those, as well as the qualification levels descriptors in tertiary education (columns 1, 2, 3).

Considering the significance of the learning outcomes' generic descriptors, there were established the qualification levels descriptors for each of the EQF levels obtained through tertiary education. They have a high level of generality, allowing them to inform and interact with the wide diversity of qualifications and qualifications types from air transport sectoral, national and/or international qualifications systems. The qualification levels descriptors will be used to analyse and to describe each professional competence of a given qualification using a tool proposed and developed within KAAT project, called grid 1 (KAAT 2018), which contains the name of the study field, study programme, the qualification title and level, the level descriptors of professional and transversal competences, as well as the minimum performance standards. The other developed tool, called grid 2 can be used for determining the correlation between qualification competences, their learning outcomes,

			LEVEL 5	LEVEL 6 - BACHELOR	LEVEL 7 - MASTER	LEVEL 8 - DOCTORATE
TRANSVERSAL COMPETENCES	Social and personal development	**10. Life and career management**	Assuming a personal and professional development long-term plan and affirmation of the spirit of initiative and entrepreneurship in personal development and career management	Self-control of the learning process, diagnosis of training needs, reflective analysis on own professional activity.	Take responsibility for contributing to professional knowledge and practice and/or for reviewing the strategic performance of teams	Development of creativity centred projects as the basis for self-accomplishment
		9. Social interaction	Familiarisation with the teamwork-specific roles, group activities and with task allocation for subordinated levels of a specialized field of work	Familiarisation with the teamwork-specific roles, group activities and with task allocation for subordinated levels.	Interaction within professional groups or institutions	Capacity to organise and lead the activities of professional groups, research groups or institutions.
PROFESSIONAL	Autonomy and responsibility	**8. Responsibility**	Assumption of the full responsibility for the nature and quality of outputs in a specialized field of work or study	Take responsibility for decision-making in predictable, unpredictable work or study context	Assuming responsibility to manage and transform work or study context that are complex, predictable, unpredictable and require new strategic approaches.	Assuming responsibility and sustained commitment to the development of new ideas or processes at the forefront of work or study contexts, including research.
		7. Supervision and assistance in relation with the nature and task complexity	Exercise management and supervision in contexts of work or study activities where there is predictable work, assuming responsibility to the quality of processes and procedures	Supervision and assistance in managing complex technical or professional activities or projects	Undertaking complex and unpredictable professional tasks under autonomy and professional independence conditions.	Demonstrate substantial authority, autonomy scholarly and professional integrity in complex and unpredictable research, education and professional context.

Fig. 5 SQFAT matrix—Qualification' descriptors for EQF 5,6,7 and 8 levels (KAAT 2019; Zaharia et al. 2020)

Learning outcomes	Generic descriptors	Qualification' Levels Descriptors			
SKILLS (functional-actionable dimension)	**6. Communication**	Communication in different contexts / environments, including foreign languages and ICT-mediated, communication adapted to various public	Communication in different contexts / environments, including foreign languages and ICT-mediated, communication adapted to various public	Communication in different contexts / environments, including foreign languages and ICT-mediated, communication adapted to various public	Communication in different contexts / environments, including foreign languages and ICT mediated, communication adapted to various public
	5. Creativity and innovation	Solving problems of work or study in a specialized field, possibly developing creative approaches, preparing technical documents and progress reports.	Development of professional and/or research projects using well known principles, methods and software within the field	Development of professional and/ or research projects integrating a wide range of methods in different fields in an innovative means.	Design and undertake original research, based on advanced methods leading to the development of scientific and technological knowledge and/or of the research methodologies
	4. Critical and constructive reflection	Prompt notification of failure to use equipment, measuring and control devices and regulations specific to a specialized field of work or study.	Adequate use of standard assessment criteria and methods to appraise the quality, merits and limitations of processes, programmes, projects, concepts, methods and theories	Pertinent and appropriate use of qualitative and quantitative assessment criteria and methods to formulate judgements and fundament constructive decisions	Critical/constructive assessment of projects and scientific research results, appraisal of the stage of theoretical and methodological knowledge; identification of knowledge and application priorities with in the field
	3. Application, transfer and problem solving	Execution of complex tasks within a specialized field of work or study, using technical documentation and tools for measuring / monitoring technological processes in normal new or changing conditions.	Application of advanced principles and methods to solve complex and unpredictable problems/situations that are typical to the field of work /study.	Integrated use of the conceptual and methodological apparatus in situation of with incomplete information in order to solve new theoretical and practical problems	Select and use advanced principles, theories and methods of knowledge, transfer of methods from one field to another, interdisciplinary approaches to solve new and complex theoretical and practical problems
KNOWLEDGE (cognitive dimension)	**2. Explanation and interpretation**	Use adequate documentation, catalogs and standards for description and integration of the principles, norms, processes in a specialized field of work / study.	Use of advanced knowledge to explain and interpret various types of concepts, situations, processes, projects etc. related to the field	Use of highly specialized knowledge in order to explain and interpret new situations in wider contexts associated to the respective field	Use advanced principles and methods to explain and interpret, from multiple perspectives, new and complex theoretical and practical problems that are specific to the respective field
	1. Knowledge, understanding and use of specific language	Use of the concepts, principles, processes and standards / regulations particular to a specialized field of work or study.	Knowledge and understanding of advanced concepts, theories and methods in the field and the specialization area;	In-depth knowledge of a specialization area and, within it, of the programme specific theoretical, methodological and practical developments;	Systematic, advanced knowledge of concepts, research methods, controversies, new hypothesis specific to the field; communication with specialists from related fields

Fig. 5 (continued)

study program and credits allocated in order to establish a certain study plan (Zaharia et al. 2020).

Based on a given qualification analysis and description realised with these three tools, the training provider will be able to elaborate the study/training plan and the disciplines syllabus adapted to the new requirements of air transport sector.

3.2 Methodology for Recognition of Prior Learning and Work Experience in Aviation

Lifelong learning paradigm values all kinds of learning—formal, non-formal and informal. Recognition of prior learning or work experience and validation of non-formal and informal learning form a cornerstone in the lifelong learning strategy. The essential tools in recognition are the learning outcomes and the description of qualifications by Grid 1 (KAAT 2020). Learning outcomes should be recognised and valued, regardless of where and how they are achieved. Such recognition of non-formal and informal learning enables learners to identify their starting point, gain entry to a programme of learning at a certain level, achieve credits towards a qualification and/or achieve a full qualification based on competences. It serves to motivate reluctant participants, add value to prior learning and save time and money by reducing or eliminating the need to relearn what has already been learned. Similarly, it enables society to benefit from skills acquired at no public cost (Commission of the European Communities 2006). In aviation to non-formal learning we associate work experience which is essential in professional pathway.

Challenges that the EU countries are currently facing in terms of rising youth unemployment rates, shortcomings of competences relevant to the labour market and increasingly elderly populations bring about non-formal and informal learning as a means of unlocking significant reserves of underdeveloped human capital. Along these lines, in December 2012, the Council of the European Union issued a recommendation recognizing the importance of non-formal and informal learning pathways to tackle the problems of specific target groups, including young people, the unemployed and unskilled people (The Council of the European Union 2012). RPL could play a key role for workers who have already become redundant or may become redundant in the future. With the help of RPL, their competencies become visible through certification, which would serve as a basis for building new career opportunities through further education or even through specific job training (KAAT 2020).

For the recognition of prior learning (RPL) in aviation to obtain a certain qualification certification, we proposed a conceptual tool, a matrix for RPL (Table 1). The situation of RPL for a certain qualification in aviation can then described more in depth.

In a rapidly changing world, it is useful and important for individuals to acquire competencies through some form of learning in order to remain employable and to face challenges they could not have previously anticipated. Formal qualification systems not always support that possibility because NQFs are mostly focused on formal learning acquired in educational institutions. The idea of formalizing knowledge, skills and competencies acquired outside the formal education system or accumulated achieved by work experience and making learning 'visible' is the key value of recognition (including evaluation) of non-formal and informal learning, or RPL.

Table 1 Recognition of Prior Learning in aviation sector Matrix (KAAT 2020)

Name of qualification		Levels of qualification recognised by RPL		
Context space of recognition		a. A lower level of qualification	b. The same level of qualification	c. A higher level of qualification
A. Organizational	1. Unregulated	A1.a	A1.b	A1.c
	2. Regulated	A2.a	A2.b	A2.c
B. Geographic	1. National	B1.a	B1.b	B1.c
	2. European	B2.a	B2.b	B2.c
	3. International	B3.a	B3.b	B3.c

3.3 *Useful on-Line Tools and Platforms for Information, Guidance and Job Offers*

One of the objectives of our research, briefly presented in this paper, is to provide structured knowledge on activities, methods and tools supporting the successful integration of information and communication technologies (ICT) and labour market information (LMI) in career guidance services and presenting potential for transferability in air transport. It aims to support air transport managers in identifying successful ICT and LMI initiatives to transfer to their context and providing guidelines on how to integrate them into their national context by identifying and analysing their essential components.

This section can serve as a starting point, as more elaborated information is available on the websites mentioned below. The ICT and LMI practices presented have been selected using a multidimensional decision framework, with a view for meeting the following quality conditions: (a) compatibility with air transport priorities for career guidance; (b) innovativeness, exhibiting sufficient and reliable evidence for positive impact and successful implementation in the framework of career guidance centres at national and/or regional level; (c) relevance for air transport qualifications and aviation labour market; (d) transferability and adaptability to other contexts; € size of training networks (KAAT 2019).

The information from our research is synthesized in Table 2.

3.4 *Enate*

In the mid and long term, the communication and career development will be continued and sustained by a European Network for Aviation Training and Education (ENATE), which will be set up starting with middle 2020, as a result of KAAT project. This network will contribute to the international dimension of the educational program in aviation and will impact the European employability in air transport sector, by producing modern, dynamic, committed and professional environment

Table 2 Platforms and tools for guidance and career development in air transport

Name	Main purpose			Web address
Developed by	Information and guidance	Education/Trainings	Job offers	
ESCO portal/EC	– Provides descriptions of 2942 occupations, 10,583 competences; – Available on 27 languages; free – Analysis of labour market data on skills and occupations – Integration with a significant number of digital platforms	– Suggest the most relevant trainings to people who want to reskill or upskill	– Connects labour market with education and training systems – Supports job mobility across Europe – Suggest the most relevant jobs to jobseekers based on their skills or – Several domains including air transport	www.ec.europa.eu/esco/portal
Civil Aviation Training Solutions (CAE)	– Has the largest civil aviation network in the world – The recruitment consultants offer a personalised service, working with candidates to find the ideal job based on experience and lifestyle	– More than 50 training locations across the globe – More than 250 + full-flight simulators – More than 2000 highly skilled instructors – Training for: business pilot training, maintenance training, cabin crew training, airlines and Fleet Operators, XR Series	– Provides recruitment services for airline pilots and aviation professionals	www.cae.com/civil-aviation

(continued)

Table 2 (continued)

Name	Main purpose			Web address
Developed by	Information and guidance	Education/Trainings	Job offers	
Groupement des Industries Françaises Aéronautiques et Spatiales (GIFAS)	– Represents the French aerospace sector at national and international level – EDEC framework (Commitment to Developing Employment and Skills)	– Technical training courses – "Aéro Fab Emploi"	– On-line platform for jobs: – Production, R & D Engineering office, Support function, Tests, Simulations, Maintenance Repair—After-Sales, Program Management	www.gifas.asso.fr/en/

(continued)

Table 2 (continued)

Name	Main purpose			Web address
Developed by	Information and guidance	Education/Trainings	Job offers	
International Civil Aviation Organization (ICAO)/UN	– Provides civil aviation Standards and Recommended Practices (SARPs) and policies regulations – Various and complex information resources on air transport sector (publications, reports, databases, surveys, forecasting etc.)	Provides training through GAT for: Aerodromes; Training Competency Development; Air Navigation Services; Security and Facilitation; Air Transport; Environment; Flight Safety and Safety Management	– ICAO Website/Careers – e-recruitment system	www.icao.int

(continued)

Table 2 (continued)

Name	Main purpose			Web address
Developed by	Information and guidance	Education/Trainings	Job offers	
Airports Council International (ACI)	– Develops standards, policies and recommended practices for airports – Regularly provides airport data and statistics, publications, reports, infographics – Information about training opportunities around the world	– ACI Global Training - the world's leading provider of airport management and operations education – Course categories: customer experience, economics, environment, management, operational, technical, safety, cybersecurity, on-line learning	– Airport Jobs on official website – Employers can send job openings to be disseminated	https://aci.aero/

(continued)

Table 2 (continued)

Name	Main purpose			Web address
Developed by	Information and guidance	Education/Trainings	Job offers	
European Transport Workers' Federation (ETF)	– Represent the social and economic interests of workers in transport (including aviation) – Ensures sectoral social dialogue in civil aviation – Provides information resources on air transport sector	– Produces training packages – Provides training materials on website		www.etf-europe.org

(continued)

Table 2 (continued)

Name	Main purpose			Web address
Developed by	Information and guidance	Education/Trainings	Job offers	
EUROSTAT/EC	– Provides high quality statistics for Europe, including for air transport sector – Free access – Data accessible on our different visualisation tools	– Provides assistance to all types of users	– Offers challenging and rewarding careers in the field of European statistics – Selection by EPSO – Job opportunities: Permanent officials and Temporary agents; Contract Agent staff; National experts; traineeships	www.ec.europa.eu/eurostat www.epso.europa.eu
www.study.com	– A personalized learning platform that is helping over 30 million people a month	– Provides fast, efficient micro learning, any subject, anytime; affordable; personalized – Over 4100 video courses, grouped by domains, education levels and goals	– Website/Career section	www.study.com

(continued)

Table 2 (continued)

Name	Main purpose			Web address
Developed by	Information and guidance	Education/Trainings	Job offers	
Academiccourses.com	– On-line platform with resources – Connects students with educators – Multilingual	– Provides general courses, preparatory years, short programs, MBA, PhD, certificates, diplomas, summer courses, on-line courses etc. – For aviation, 211 courses; e.g. aeronautical engineering, air traffic control, aviation management, safety, cabin crew, helicopter, pilot, aircraft maintenance	–	www.academiccourses.com

(continued)

Table 2 (continued)

Name	Main purpose			Web address
Developed by	Information and guidance	Education/Trainings	Job offers	
Airsight	– Innovative consulting services based on in-depth knowledge for various aviation aspects	– Training for airports, air navigation service providers, for civil aviation authorities and organisations – In-house courses (Berlin) & virtual class – Trainings are live and instructor-led with interactive participation – English and German	– Email to jobs@airsight.de – For student jobs, internships, Master/Bachelor/PhD Thesis, -email to studentenjobs@airsight.de	www.airsight.de/
Jobsinaviation.com	– Articles and newsletters	– Assist through its Aviation Training Directory for: engineer, pilot, cabin crew and aerospace training courses	– 108 job offers in aviation (3rd of May 2020): aircraft engineering, aerospace engineering, pilots, flight attendants, airports, airlines	www.jobaviation.com

(continued)

Table 2 (continued)

Name	Main purpose			Web address
Developed by	Information and guidance	Education/Trainings	Job offers	
Aviationjobsearch.com	– Provides news from aviation sector – English and German	– Provide several training providers' information, though on-line platform	– 345 job offers in numerous categories of aviation sector (3rd of May 2020) – On-line register and application process	www.aviationjobsearch.com www.aviationcourses.com

inside project organisations thanks to the sharing and integration of good practices and new up-to-date subjects into educational programmes and initiatives.

The main objectives of network are to ensure and enhance on the long term a systematic university-business dialog for a coherent education and training in aviation according to the demands of the labour market and to disseminate relevant results of research.

ENATE will be an interlocutor for EASA, ICAO, ACI, European Commission (DG Transports, DG Education and DG Employment) on the education and training in aviation and furthermore for ensuring a right proposal for a better match between the need of occupations in terms of competences and the qualifications provided by HEIs or training providers. The Association will be a permanent barometer for ensuring the trio new jobs – new skills – new study programmes. The association will have several branches in partner countries, registered according to the local legislation.

4 Conclusion

Based on the analysis of air transport and the developed reports we can conclude that the air transport industry is confronting with important requirements skilled personnel and with an important shortage in training capacity.

Within our research, we have identified skills shortage for several occupations in this sector. In aviation, compared to industries, including other transportation activities (naval, rail, automobile), many occupations require specific skills and companies report strong difficulties in finding qualified professionals with experience in the air transport industry. In addition, the lack of attractiveness of the technical and even scientific study programmes and of industrial sectors stresses recruitment tensions. For example, the main difficulties are reported for professions which require more than one specialisation: software aeronautical architect, computational engineer with aeronautical knowledge, environmental air transport engineer etc. There are also classical occupations in aviation for which it is difficult to find people with the adequate qualifications, as for example mechanic for aircraft maintenance and for airline pilots. For this reason, many companies from air transport industry have opened new trainings for covering the shortage of qualifications.

One of the resulted conclusions is that the recruitment levels on most occupations in aviation tend to change. The degree levels required for aviation occupations have tended to increase, more occupations are regulated or require new skills and knowledge. The increase in the technical nature of the air transport operations have an impact on recruitment levels, on nature of skills and knowledge and thus increase the share of engineers and managers. New knowledge and skills are required as for example digital skills, renewable energies, environmental protection etc.

There are many jobs and emerging jobs which requires interdisciplinary competences and, thus, interdisciplinary qualifications. Currently, there are not enough study programs to cover this need for interdisciplinary competences as for example:

IT and aviation, economics and aviation, environment and aviation, law and aviation. Considering these new jobs, new knowledge and necessary skills for the digitalisation of aviation, the education and trainer providers must ensure appropriate curricula and furthermore curricula need to undergo chances on the new aviation labour market.

To add, more than one out of two jobs will not be the same in ten years. The aviation digitalization makes existing skills to be outdated. In most situations, the qualifications have not evolved with the same rhythm needed by enterprises. Automation and the digital interfaces have been multiplying, triggering the need of new digital skills. For solving the current situation higher education institutions or aviation bodies have initiated new interdisciplinary study programs and trainings.

All the instruments presented in this paper have the role of informing and guiding, of providing specialized training solutions or to facilitate the relation between stakeholders, thus leading to the career development of staff, whether they are graduates or people already in the field of work. Through study programs or trainings, learning outcomes are being acquired and websites can serve as connecting bridges. Ensuring new skills for the new occupations is the main challenge for education and training in aviation for the next period. Moreover, given the unprecedented crisis currently affecting the world, there is a need to reconfigure processes and activities in all areas, and especially in air transport, noting the urgent need to use online technologies.

Acknowledgements This research was co-funded by the Erasmus + programme of the European Union, through the project "Knowledge Alliance in Air Transport (KAAT)", project no. 588060-EPP-1-2017-1-RO-EPPKA2-KA.

References

Airports Council International (ACI)-Europe (2020a) Impact of COVID-19 on European airport passenger traffic. Brussels, Belgium. Available from https://www.aci-europe.org/airport-traffic-covid-19/. Accessed 20th Apr 2020

Airports Council International (ACI) (2020b) ACI world data shows dramatic impact of COVID-19 on airports. In: ACI media releases (3rd of July 2020). Available from https://aci.aero/news/2020/07/03/aci-world-data-shows-dramatic-impact-of-covid-19-on-airports/. Accessed 14th of July 2020

Airports Council International (ACI)-Europe (2020c) COVID-19 and airports. Passenger traffic and revenue impact. Updated Forecast (July). Available from www.aci-europe.org. Accessed 14th of July 2020

Commission of the European Communities (2006) Communication from the commission adult learning: It is never too late to learn, COM 614 Final. Belgium, Brussels

IATA (2020) Industry Losses to Top $84 Billion in 2020. Available from https://www.iata.org/en/pressroom/pr/2020-06-09-01/. Accessed 3rd of July 2020

ICAO (2019) Aviation benefits report 2019. Available from https://www.icao.int/sustainability/Documents/AVIATION-BENEFITS-2019-web.pdf. Accessed 15th of March 2020

ICAO (2020) Economic impacts of COVID-19 on civil aviation. economic development of air transport 2020. Available from https://www.icao.int/sustainability/Pages/Economic-Impacts-of-COVID-19.aspx. Accessed 29th of April 2020

Knowledge Alliance in Air Transport (2018) Report on occupational analysis in air transport industry. Available from: http://www.kaat.upb.ro/wp-content/uploads/2018/12/KAAT_WP1_R1.1_Report-on-occupational-analysis_Final.pdf. Accessed 29th of June 2020

Knowledge Alliance in Air Transport KAAT (2019) Report on qualifications in air transport industry. Unpublished paper. Available from www.kaat.upb.ro/publications. Accessed 24th of February 2020

Knowledge Alliance in Air Transport KAAT (2020) Methodology for recognition of prior learning and work experience in aviation. Unpublished paper. Available from www.kaat.upb.ro/. Accessed 28th of April 2020

Pastore, F. (2018) Why is youth unemployment so high and different across countries? IZA World of Labour Evidence-based policy. Available from https://wol.iza.org/articles/why-is-youth-unemployment-so-high-and-different-across-countries/long. Accessed 13th of July 2020

Pearce B. (2020) COVID-19 Wider economic impact from air transport collapse. In: IATA economics, 7th of April 2020. p. 10. Available from https://www.iata.org/en/iata-repository/publications/economic-reports/covid-19-wider-economic-impact-from-air-transport-collapse/

Statistica Research Department (2020) Coronavirus: airport passenger traffic change in Europe by region 2020. In: Transport and logistics. Aviation. 27th of April 2020. Available from https://www.statista.com/statistics/1107051/coronavirus-impact-airport-passenger-traffic-europe-region/. Accessed 29th of April 2020

The Council of the European Union (2012) Council recommendation of 20 December 2012 on the validation of non-formal and informal learning. In: Official journal of the European union: 2012/C 398/01. Brussels, Belgium. Available from https://eur-lex.europa.eu/legal-content/EN/TXT/?uri=celex:32012H1222(01)

Zaharia SE, Pavel AP, Hirceag C (2018) Better partnerships for better skills and employability in air transport. In: EDULEARN18 proceedings: 10977-10987. https://doi.org/10.21125/edulearn.2018.2707. 10th International Conference on Education and New Learning Technologies, 2-4 July, 2018, Palma de Mallorca, Spain

Zaharia SE, Toma S, Boc RL (2020) Sectoral qualifications framework for air transport industry. In: EDULEARN20 proceedings, international conference EduLearn. 6-8 of July 2020; Palma de Mallorca, Spain. Available from https://iated.org/concrete3/paper_detail.php?paper_id=81923

Drivers of Change for Smart Occupations and Qualifications in Aviation

Sorin Eugen Zaharia, Casandra Venera Pietreanu, and Adina Petruta Pavel

Abstract In the context of aviation education management, the necessary conditions for developing emerging jobs and improving training must take into consideration drivers of change that will transform the air transport system and require the development of new skills and knowledge. The paper focuses on the challenges and changes in future occupations and identifies new qualifications needed to meet the trends in air transport. The changing nature of work in following years considers both the occupations that are going to disappear and those that will be created. These will imply important modifications in labour market, education and training. After analyzing the perspective of the labour market requirements and the aspects that have the potential to transform businesses or even destabilize the industry, future occupations are reviewed in terms of economic benefit, new modes of consumption and influences on the market. Since the opinions of specialists regarding the forecasts for the opportunities and threats in aviation vary, the authors suggest a shift to a knowledge-based economy supporting the changing nature of work in the need for high qualified employees. Thus, the goal to provide the labour market with competent workforce through cooperation between governments, education institutions, industry employers and regulators is supported.

Keywords Aviation future jobs · Labour market · Smart qualifications · Smart occupations · Societal responsibility

S. E. Zaharia · C. V. Pietreanu (✉)
Faculty of Aerospace Engineering, University Politehnica of Bucharest, Polizu Street 1-7, Sector 1, 011061 Bucharest, Romania
e-mail: casandra.pietreanu@yahoo.com

S. E. Zaharia
e-mail: sorin.zaharia@gmail.com

A. P. Pavel
University Politehnica of Bucharest, UNESCO Chair "Engineering for Society", Spl. Independentei 313, 060042 Bucharest, Romania
e-mail: adinappavel@gmail.com

M. Petrović and L. Novačko (eds.), *Transformation of Transportation*, EcoProduction,
https://doi.org/10.1007/978-3-030-66464-0_13

1 Introduction

As a consequence of the development of airport digitalization, forward-looking thinking on innovative environmental improvements and societal responsibility which can enhance a sustainable development in aviation, new occupations are arising. Technological, socio-economic or geopolitical factors are still solid pillars that can affect the industry, but important concerns of future imply the dynamics of societal responsibility, data protection, surveillance and decision-making. In this context, the research mirrors a proposal of new qualifications for 'smart' occupations, which will be described in terms of the necessary learning outcomes demanded by the evolution of air transport.

After analyzing the wide range of qualifications required in aviation, the authors outline the need for highly qualified employees and analyze the characteristics of the labour market in air transport, future jobs and emphasize on the drivers of change regarding emerging occupations.

The methodology of the research combines analysis of the requirements and recruitment needs of various employers: airlines, airports, handling companies, regulatory bodies, etc. On the other hand, since improved skills for new occupations represent the main challenge for education and training in aviation, discussions with universities and training providers were carried. All these studies were performed in compliance with the European and national Qualifications Framework and international regulatory bodies: European Aviation Safety Agency (EASA), International Air Transport Association (IATA) and International Civil Aviation Organization (ICAO), etc. More so, questionnaires on new occupations and qualifications were developed by the authors in order to collect feedback from various employees in the aviation sector. The questionnaires provided important information about emerging occupations and future workforce, and outlined greater collaboration opportunities between the industry and educational institutions. Thus, the current research is based on a wide documentation in order to identify various factors and forces called 'drivers of change'.

2 Characteristics and Development of the Labour Market in Air Transport

2.1 Context

A wide range of operations are nowadays affected by the COVID-19 health emergency, so aviation has lost for the moment its important role in contributing to economic growth (by supporting employment, enabling world trade, developing tourism, public funds and investment increase, etc.) or social benefits (increased availability/access in all areas of the world, consumer welfare, humanitarian aids/support or sustainable development). In the context of air transport growth before the

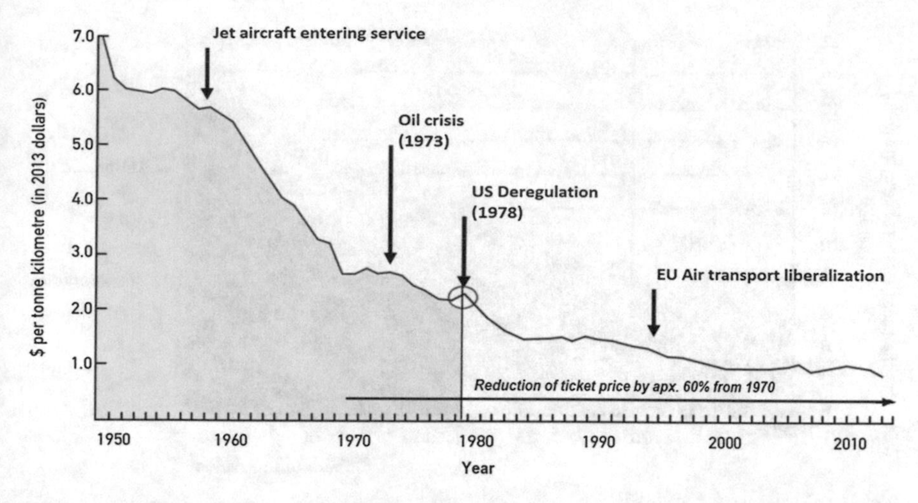

Fig. 1 Evolution of average price of air travel [in $ per RTK (revenue tonne km)] (Reproduction from IHLG 2017)

pandemics (in 2019, 4.5 billion passengers were carried by air (Mazareanu 2020), the number of employees were also increasing, transforming the labour market. A key driver in the growth of the number of passengers worldwide has been the steady decrease in the real cost of air travel. Since 1970, the cost of a ticket has been reduced by over 60% (Fig. 1).

Furthermore, through deregulation of the US aviation market in the 1978 and in the European market in 1986, the development of fuel and cost efficient aerospace technologies or the introduction of LCC, more and more people afford to travel by air. Studies show that affordability (i.e. airfare, ancillaries and taxes) ranks first regarding personal reasons in terms of choosing transport characteristics (Airlines for America 2018). Even business travellers value affordability; in 2017, 38% of passengers reported flying at least once from an airport that wasn't the closest to their home or office (Airlines for America 2018).

The growth of passenger traffic (Fig. 2.) holds important economic potential which supports all states in achieving the United Nations 2030 Agenda for sustainable development.

In this context, since the global demand for air transport was meant to increase continuously by an average of 4.3% per year over the next 20 years (i.e. by a factor of 2.3); by 2034, this growth path would have contributed to (ATAG 2018): 15.5 million direct jobs and $1.5 trillion of GDP to the world economy and a total of 97.8 million jobs and $5.7 trillion in GDP once the impacts of global tourism were considered (for an OPEN SKIES Liberalised Scenario) (Fig. 3).

However, due to COVID-19 outbreak, industry-wide impact shows 70% fall during Q2 and 38% fall in average in 2020 in revenue passenger killometer (RPKs) and an unprecedented revenue loss of $252 bn (IATA, Airlines Financial Monitor 2020) (see Fig. 4).

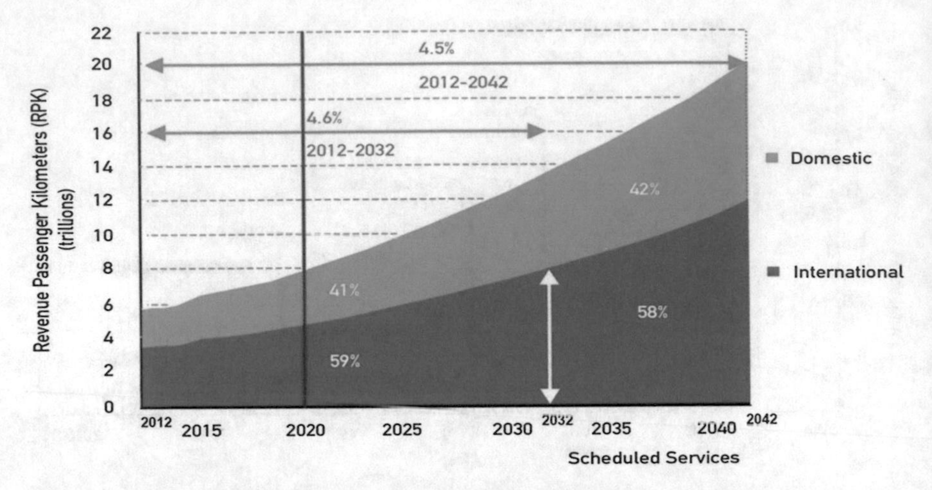

Fig. 2 Total passenger traffic: history and forecast (Reproduction from IHLG 2017)

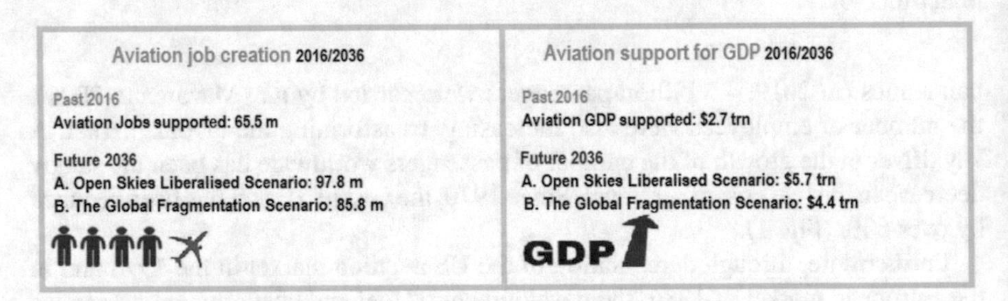

Fig. 3 An assessment of the next 20 years of aviation (Two Scenarios) (Reproduction from ATAG 2018)

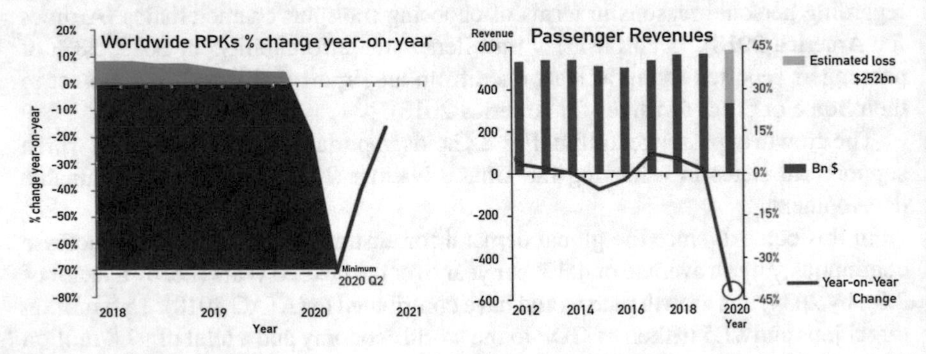

Fig. 4 Total passenger traffic: history and forecast (Reproduction from IATA 2020)

All these aspects reflect in the dynamics of the labour market. Hereinafter, the labour market in air transport will be analyzed in the context of the international regulations with respect to required qualifications, creation of new training institutions or expansion of existing ones, and also the recognition of prior learning.

2.2 *Characteristics of the Labour Market in Air Transport*

Air transport, as an important source of economic activity, creates jobs that serve all actors in aviation: airlines, airports and ATS, etc.; and includes many categories of occupations. Besides operational aspects, one should consider the importance of the manufacturing sector which is also responsible for technological improvement. Out of 65.5 million jobs supported by aviation, 10.2 million are direct jobs (airport operators and other airport base-roles, airline related, engineers, ATC, executives), 10.8 million jobs are supported indirectly and include suppliers, manufacturers and business support roles, and the rest are induced jobs (7.8 million) and catalytic (36.7 million) (ATAG 2018).

The dynamics of aviation business models and the technology progress will have strong impact on the employment over the coming years. The human resource in aviation must be adaptable and highly qualified in order to work in a complex, highly competitive, interdisciplinary and multicultural environment. Thus, employees must be capable to adapt in a dynamic field that is characterized by increasing pressure. But regardless of the human factor, the drivers of change in the labour market range from creating new jobs to eliminating the ones that will no longer be needed. With respect to the above mentioned, the labour market in air transport, the education and training available in the context of the European regulation in terms of qualifications will be analyzed thoroughly hereinafter.

After studying ICAO and EC documents on qualifications and focused groups with stakeholders, one conclusion is that nowadays there is a strong growth in recruitment needs, in many cases for newer occupations, but there is also a lack of competent profiles, especially for interdisciplinary ones (ICT and aviation, environment and aviation, marketing and aviation, etc.). A 20 year forecast for pilots, maintenance personnel and ATC developed by ICAO indicates that since maintenance employees will increase their numbers by a 3.9 factor by 2030, the annual training will mirror a shortage of 8352 people (ICAO 2011). This is combined with the fact that maintenance personnel will have to be high-calibre graduates. So, the future will show important requirement for pilots and maintenance personnel, but will face significant shortage in training capacity.

For the future, in order to predict the labour market dynamics, a correlation between service/process automation and the number of jobs that will be required in the future, must be considered. On the other hand, with respect to licensed aviation personnel, the demand for pilots, maintenance personnel and air traffic controller should be correlated to aircraft delivery plans. An ICAO Report (Global and Regional 20-year Forecast) shows that commercial air transport grew strongly from 2000 to

2010 due to the largest number of aircraft orders ever recorded and the emergence of new airline operators (ICAO 2011). This forecast is very important to enable training plans, create new training institutions or expand the existing ones, but also to find effective means for the recognition of prior learning or validation of experience for a smooth career development and a flexible work market; since through this recognition, the period of training could be shortened and the capacities could be saved (Braňka 2016).

A report of World Economic Forum shows that 65% of children entering primary school today will have a job that doesn't even exist yet, and in just ten years from now, more than one out of two jobs will not be exactly the same, undergoing transformations (World Economic Forum 2017). This is also applicable for aviation, and outlines the rationale for the importance of the ability to anticipate future employment characteristics in order to develop predictive analysis on the opportunities and threats, in order to maximize the access to future jobs.

As a conclusion, the optimization of aviation labour market generates substantial economic gains in the face of traffic projections, indirectly improves educational and training programmes, and supports better career and salary prospects of future employees (thus, the quality of life). The wellbeing related to workplace takes into consideration employees' satisfaction, health and professional development, aspects that can strengthen of an organizational performance; so this is a win–win situation. For this, advisable measures include lifelong learning, empowerment and self-development of employees, improved information transfer throughout the company, greater work force diversity and partnerships, balanced earnings and career perspective development for women, and preoccupation with respect to employability and job security (European Commission 2019). These aspects are components of societal responsibility.

2.3 Aviation Labor Market in the Context of Digitalization

In 2016, 52% of employees in European Union presented moderate ICT level, which is needed to carry out their job tasks, while 19% required a basic level (CEDEFOP 2016). Digital transformation creates new types of jobs and since the EU labour market has high demand for advanced digital skills, the challenge is to manage the impact by developing this skills and providing flexible workforce. Several concepts like "collaborative decision-making", "self-services", "digital management", "operations centre", related to the digital transformation of airports are evolving and are of interest for both passenger satisfaction and for internal processes of air transport (Fig. 5).

The application of IT is also important for the optimisation and modernisation of operations/procedures, e.g.: airport collaborative decission making (ACDM), airport operations center (APOC), ATM (air traffic management), CRM (crew resource management), for developing Airport 4.0. More so, programs like SESAR based on IT deployment in air transport creates the need for new interdisciplinary study

Fig. 5 Examples of smart airport applications

programmes in higher education. Before Airport 4.0, partially due to the airport's traditional business to business (B2B) model, airports were not at the top of the digital rankings (compared to other industries). But nowadays, airport digital transformation is intensifying operational efficiency and passenger experience (airport operations, passenger journey, ancillary revenues) (Zaharia and Pietreanu 2018).

In a nutshell, the future of the aviation labour market will provide fewer jobs involving physical, routine or repetitive tasks; while occupations related to ICT (data science, machine/deep learning, robotics, artificial intelligence, etc.) will gain ground. More so, studies show that due to Covid-19 pandemics reflected in prolonged economic shutdown, the labour trend that considers digitalization will be accelerated, since algorithms and robots can't get sick and automation is more rapidly adopted during economic decline (Connley et al. 2020). Surely, this must be combined with a substantial up-skilling.

It is a fact that digital technologies in aviation will require the evolution or reshaping of current jobs. But more than 95% of job offers in aviation digitalization field also mention required profile skills that do not fall under technical know-how, such as: creativity, autonomy or good knowledge of foreign languages (soft skills), or management (transversal skills). Thus, the hard part of creating new types of jobs demanded by digitalization is to anticipate future skills requirements in order to be prepared, then share a common expertise and create a safety net for employees (Zaharia and Pietreanu 2018). One must keep in mind that technological advancement has always led to increased prosperity and productivity.

By evaluating the future labour market from the perspective of digitalization, 'greenization' and sustainable development, the partnerships between governments, higher education institutes and training providers could be strengthened in order to better manage the transformative impact on future aviation development. As automation and the digital interfaces develop, setting the need for new digital skills is mandatory. However, current employees do not master these skills; consequently, as a solution for this situation, educational institutions or aviation bodies initiate new interdisciplinary study programs and on job trainings (Sect. 3.3).

3 Smart Qualifications for Smart Occupations. Drivers of Change

Over the last years, several terms from the aviation industry have been introduced in current speech and usage: "smart technologies", "smart operations", "smart buildings", "smart airports", etc. They are all related to technological improvements and digital transformation entering the industry. So, taking into consideration these ongoing working conditions, aviation jobs should be performed by "smart employees", having "smart occupations".

The ever changing nature of the jobs in aviation has an important technological pillar, and in the background must be supported by study programs that provide smart interdisciplinary qualifications. This study on smart occupations researches current and future skill sets and recruitment patterns in an industry that must be adaptable and not merely speculate on future risks and opportunities (World Economic Forum 2016).

New types of occupations in aviation need to be created, and the drivers of change involve social, economic and environmental factors and the regulations designed to address these factors (IATA 2018). The analysis of future trends aims to envision the way the future might unfold, taking into consideration new modes of consumption, tensions between data privacy and surveillance, cybersecurity, terrorism, strength and volatility of global economy, geopolitical instability, or even infectious disease and pandemics (IATA 2018).

The technological revolution in aviation and the environmental constraints favour the emergence of new occupations which demand new qualifications (defined by new skills and knowledge), in many situations interdisciplinary and with a very high level. After clarifying the long term qualification policy in air transport, this chapter will set the grounds for the description of the emerging smart occupations, with a particular approach on the skills and qualifications needed for digitalization. The authors will identify and describe new qualifications that meet the development trends, but also outline the profile of the human resource that will be implicated in this area.

3.1 Societal Responsibility

Besides employees' work engagement and career satisfaction (Ilkhanizadeh and Karatepe 2017), societal responsibility in aviation presumes equity–social concerns, ecological–environmental considerations, economic mergers, community involvement and ethical marketing (Elkington 1998) that a company should consider in its business. Societal responsibility is cross sectoral, and sometimes might require partnerships with other organizations. It represents a balance between corporate rights and obligations (provided by the needs and expectations of its stakeholders), this way considering social, economic and environmental aspects.

Nowadays, the aviation labour market does not define or provide specific occupations in this field, nor the competencies necessary for a person to develop and implement societal responsibility actions are outlined. But, surely, employees involved in such activities need to have knowledge and skills related to:

- Sustainability
- Communication
- Environment
- Energy issues.

KLM Royal Dutch Airlines has analyzed how organizational culture affects employees and their overall performance, so described necessary actions for achieving proper results and developed a strategic plan for societal responsibility within the group. According to the plan, societal responsibility strategy should be founded on 4 pillars that consider responsible human resources, customer experience, environment and local development (Air France KLM 2014) (Fig. 6).

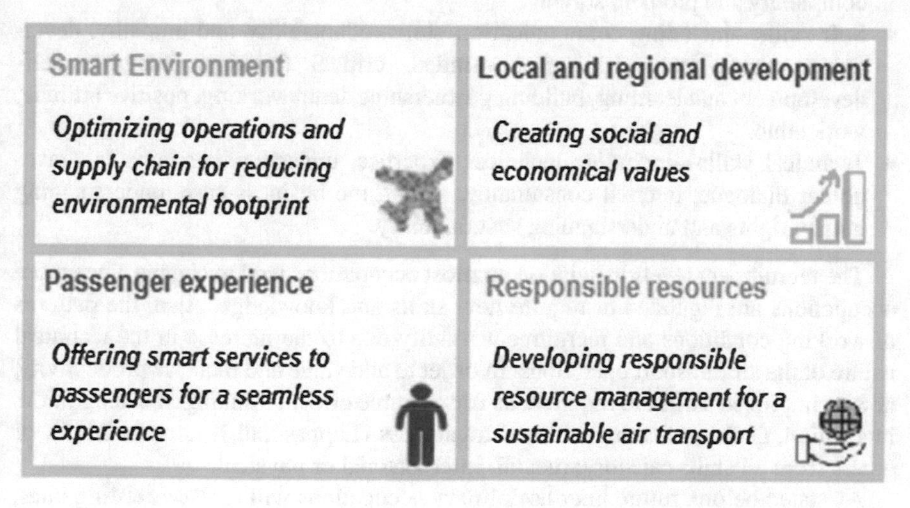

Fig. 6 Corporate societal responsibility pillars

Although it is difficult to clarify what skills and knowledge would be required for a societal responsibility role within an organization due to the interdisciplinarity of the issue, based on interviews with stakeholders, the authors have drawn the conclusion that the key competencies for employees which will prepare and implement societal responsibility actions include:

- Teamwork communication
- Systemic thinking (connecting the dots)
- Understanding and dealing with business situations (i.e. business acumen), and
- Aviation field knowledge (for better comprehension of current and future issues in air transport industry).

Given the importance of societal responsibility in aviation, it is imperative to have qualified and well-trained employees working or leading these actions for a strong and balanced growth.

3.2 The Air Transport Industry—Relying on Higher Skill Levels

The author's survey on new occupations in air transport indicates that digital skills are crucial for innovation in air transport and for supporting the trends in labour market. In this case, the skills become interdisciplinary, and can be classified as such:

- Business/Professional skills—including building insight, decision making, commercial awareness, IT, innovation, strategic awareness, leadership, handling complexity and problem solving.
- Soft skills—including communication skills, adaptability and empathy, developing others, teamwork, open minded, critical thinking, integrity, self-development and learning, building partnerships, team working, positive attitude, work ethic.
- Technical skills—including technical expertise, understanding impacts, stakeholder dialogue, internal consultancy, selling the business case, understanding human rights and understanding sustainability.

The recruitment levels in aviation on most occupations tend to change since more occupations are regulated or require new skills and knowledge. Also, the patterns of working conditions and recruitment modify due to the increase in the technical nature of the air transport operations. In order to add value and increase productivity, new or improved skills are required, as for example critical thinking, metacognition, innovation, decision making, cultural awareness (Lappas and Kourousis 2016), or others from all skills category: mental, interpersonal or physical.

As stated before, future interdisciplinary occupations will replace existing ones, so in order to fill the gap given by the lack of interdisciplinary study programmes, new curricula should provide metacompetences which in turn produce specific skills necessary in the evolutive and competitive aviation field, such as: discipline/self-discipline, entrepreneurial mentality (Miclea 2020; Zaharia et al. 2020). Also, autonomy is imperative in a digitalised environment, becoming a way to enhance the ability to exploit knowledge.

Management and human resource development within each organization should ensure the continuous improvement of the level of competence of key personnel involved, considering the interdependence between demand and supply of qualified personnel at national, regional and global levels. In aviation, continuous training is required throughout the entire career of an employee, thus also adequate assessment tools to recognize the results of different stages of training and sometimes tools for the validation of learning (or experience). To answer this need, University Politehnica of Bucharest through the medium of UNESCO Chair "Engineering for Society" is implementing the project "Knowledge Alliance in Air Transport", which intends to develop a Methodology for Sectoral Qualifications Framework and recognition of prior learning and work experience in aviation (KAAT 2020).

The air transport industry is increasingly relying on transversal competences. Thus, the higher educational programme "IT Applied in Aviation" developed by UNESCO Chair "Engineering for Society" as a result of good implementation of the project "Knowledge Alliance in Air Transport", provides students with advanced knowledge, skills and attitudes needed for their future professional activity in order to optimize the access to the ever-changing labour market. This master programme is an international one, developed by University Politehnica of Bucharest (Romania) in partnership with University of Zagreb (Croatia), Ecole des Mines d'Albi (France), University of Zilina (Slovakia), Instituto Superior Técnico from Lisbon (Portugal) and University of Strasbourg (France).

3.3 IT Applied in Aviation (ITAA) Master Programme—A Pathway to Developing Smart Qualifications

Smart occupations of the future are analyzed in order to predict possible key disruptions and their impact on recruitment patterns. Another reason is that in this case, adaptive actions should be conducted. The ease of the transition lies in anticipation and preparation; this can be achieved by deeper thinking change management.

Regarding education and training, the authors have identified 2 important aspects:

- The shortage in qualified workforce for new occupations (considering the multidisciplinary competences criteria);
- The necessity of improvement and innovation in the curricula (considering the multidisciplinary criteria for education and training).

With respect to these two recommendations, for developing the aviation labour market in the context of digitalization, University Politehnica in Bucharest, through the UNESCO Chair "Engineering for Society" proposes a master program: "IT Applied in Aviation (ITAA)". The rationale is that the specialists in ICT and the ones in aviation don't have the same background and don't share a common expertise. Thus, this cross-sectoral program represents an innovative approach, integrating various study modes: modular/personalized/mobile learning, using virtual/open educational resources and platforms. The modules are: ICT modules or Air transport specific.

In class lectures or e-learning, on-site visits and projects completed with the help of industry partners suppose that all future graduates will have in-depth knowledge in applying information and communications technology in air transport. ITAA students interaction within professional groups or institutions imply performing activities with the exertion of roles specific to teamwork on different hierarchical levels and assuming leadership roles, promoting the spirit of initiative, dialogue, cooperation, positive attitude and respect for others, diversity and multiculturalism, and continually improving the students own activities.

For the purpose of insertion of ITAA students into the labour market and adaptation to the dynamics of its requirements, besides knowledge of information technology and communication applied in aviation, social, personal and professional development, self-control of learning and efficient use of language skills are targeted. This way, the student can become a manager of his own continuous training.

The competences of the master program IT Applied in Aviation include:

- Conducting calculations and demonstrations for solving ICT tasks for aeronautical engineering;
- Selection and application of principles and methods for modelling ICT processes in aeronautics;
- Integrated use of software applications to solve tasks specific to aeronautical activities;
- Managing processes specific for IT with applications in the aeronautical field;
- Implementation of data processing algorithms for designing optimized aeronautical processes;
- Developing innovative IT products with aerospace applications.

The knowledge gained at IT Applied in Aviation master program regards implementing widely and more collaboratively outcomes of the assimilation of information through learning:

- Using the in-depth knowledge of specialized disciplines to explain and interpret new situations;
- Interpret the developments/performances in ICT applied in aviation;
- Interpreting optimization models for computer-aided design of products, characteristic to airports and airlines;
- Using knowledge in computer-aided design software packages to optimize the performance of aeronautical services;
- Use of expertise to explain and interpret new IT technologies used in aviation;

- Explaining and interpreting with high degree of detail the methodology of developing IT innovative products in aviation.

The learning outcomes including knowledge, skills and attitudes, have the role of anticipating and preparing for the current transition in aviation labour market. The future of smart occupations in aviation will bring specificity to skills landscape. The necessary skills that are developed within this master program involve the integrated use of conceptual and methodological criteria, including the communication in different environments. The description of possible new smart occupations which can be provided by the master IT Applied in Aviation is given in Table 1.

The above table shows a merger between ICT and aviation, but also economics, as a basis for the description of interdisciplinary occupations, their role, and the soft skills necessary for their development. Other emerging occupations could include: Chief IoT, Airport UX Designer, Growth Hacker, Virtualization Engineer, Software developer for air transport.

Table 1 Description of new occupations

Category/subcategory	Example	Role	Necessary soft skills
Aviation & IT/Big Data applied in aviation	Big Data Analyst	Responsible for creating and administrating database architecture and data modeling. Creates analysis and segmentation tools	Communication skills; Intellectual curiosity; Entrepreneurial spirit; Organization and rigor
Aviation & IT/Data storytelling	Data Storyteller	Best communicate information from the data through a story	Communication skills; Curiosity; Sense of organization; Editorial ease
Aviation & IT/Cybersecurity	Head of the ISS	Proposes to the competent authority the PSSI and ensures its application	Analytical mind; Management; Communication skills; Organization and rigor
Aviation & IT/Search Engine Optimization	SEO Manager	Optimizes a website to get the best places on the search engines	Management; Communication skills; Sense of organization; Curiosity
Aviation & IT/Data science	Data-Miner	Analyzes large quantities of data and reveals correlations between seemingly unrelated phenomena in order to anticipate trends	Analytical mind; Communication skills; Organization and rigor

(continued)

Table 1 (continued)

Category/subcategory	Example	Role	Necessary soft skills
Aviation & Economics/User Experience (UX)	UX Manager	Lead multidisciplinary teams and ensure compliance with the terms of reference of the mission	Management; Pedagogy and leadership; Team spirit and collaboration; Communication skills; Sense of organization; Curiosity
Aviation & Economics/User Interface (UI)	UI Designer	Brief graph, creation of graphic identity and declination	Analytical mind; Editorial ease; Curiosity; Organization and rigor
Aviation & Economics/E-CRM	CRM Responsible	Leads the strategic directions in CRM and the implementation of tools and processes to achieve the objectives of Customer Relationship	Pedagogy and leadership; Analytical mind; Management; Listening skills; Sense of organization; Team spirit and collaboration; Entrepreneurial spirit
Aviation & Economics/E-reputation	E-reputation Manager	Manage a brand's reputation and consistency online	Management; Pedagogy and leadership; Communication skills; Curiosity; Sense of organization

4 Conclusions

While dealing with troubled times, the evolution of future aviation is defined by uncertainty; however, an optimistic view would argue that the number of future jobs will at least be equal to the ones that will become redundant. It is unthinkable to manage a sector as air transport without using ICT technologies or without promoting environmental protection actions.

This complex system will need specialists who master interdisciplinary skills and knowledge, as for example in informatics and in aviation for developing better platforms and intelligent systems for air transport. In order to provide this, we would need to have an education and training strategy in place to properly equip, coach and educate the next generation; otherwise we might end up with lots of new jobs, but unemployed candidates that are unable to perform them.

Therefore, it is important to start by making adjustments in elementary education and some major, long-overdue upgrades in computer science instruction at the secondary level; then a coherent professional pathway should be developed through the instrumentality of a collaboration of the industry, regulatory organizations and higher education institutions. An improved forward looking qualification framework must cover the significant skills shortage/gap.

Faced with the challenges in aviation created by pandemics, resource shortage, climate change, geopolitical instability, traffic management, technological developments and others, there is a need for new ways of thinking outside the box and building the future. This must be achieved through innovation, flexible solutions, smart operations and technologies, and improved future functionality, which are the grounds of new interdisciplinary qualifications.

The research on future occupations shows that new approaches and ideas across disciplines are the grounds for smarter solutions. Regarding the above analysed drivers of change, the following aspects should be considered while reviewing the need for high qualified work force: economy, values and communities, technology and environment; thus the range of smart solutions related to the aviation industry regard air transport and IT solutions, the environment, intermodality, planning, etc. The smart solutions contribute to 25% of the revenue growth by digital innovations, but the general challenge remains attracting and retaining skilled employees and achieving their well-being, since employees are the most important and in fact the only remaining realistic challenge of competitiveness.

Acknowledgements This research was co-funded by the Erasmus+ programme of the European Union, through the project "Knowledge Alliance in Air Transport (KAAT)", project no. 588060-EPP-1-2017-1-RO-EPPKA2-KA.

References

Air France KLM (2014) Corporate social responsibility report

Airlines for America (2018) Air travellers in America. Findings of a Survey Conducted by Ipsos

ATAG (2018) Powering global economic growth, employment, trade links, tourism and support for sustainable development through air transport. Aviation Benefits Beyond Borders

Braňka J (2016) Understanding the potential impact of skills recognition systems on labour markets: research report. International labour Organization, Geneva. ISBN: 9789221313540

CEDEFOP (2016) The great divide: digitalisation and digital skill gaps in the EU workforce. ESJsurvey Insights No 9. Thessaloniki, Greece

Connley C, Hess A, Liu J (2020) 13 ways the Coronavirus pandemic could forever change the way we work. CNBC

Elkington J (1998) Cannibals with forks: triple bottom line of 21st century business. Capstone Publishing Ltd., Mankato, MN, USA

European Commission (2001) Green paper-Promoting a European framework for corporate social responsibility. eur-lex.europa. 52001DC0366

IATA (2018) Future of the airline industry 2035. International Air Transport Association

IATA (2020) Airlines Financial Monitor. IATA Economics

ICAO (2011) Global and regional 20-year forecast pilots, maintenance personnel air traffic controllers. International Civil Aviation Organization, Montreal, Canada

IHLG (2017) Aviation benefits 2017 report. Industry High Level Group

Ilkhanizadeh S, Karatepe OM (2017) An examination of the consequences of corporate social responsibility in the airline industry: work engagement, career satisfaction, and voice behaviour. J Air Transp Manage. https://doi.org/10.1016/j.jairtraman.2016.11.002

KAAT (2020) Methodology for recognition of prior learning and work experience in aviation. Knowledge Alliance in Air Transport

Lappas I, Kourousis K (2016) Anticipating the need for new skills for the future aerospace and aviation professionals. J Aerosp Technol Manag 8(2)
Mann C (2018) Digitisation of airport processes. Smith detection
Mazareanu E (2020) Air transportation—statistics and facts. https://www.statista.com/topics/1707/air-transportation/
Miclea M (2020) Cele patru metacompetente de care avem nevoie pentru ceea ce ne asteapta in viitor, Timponline
World Economic Forum (2016) The future of jobs, employment, skills and workforce strategy for the fourth industrial revolution. Global Challenge Insight Report
World Economic Forum (2017) Digital transformation initiative. Aviation, travel and tourism industry (White paper)
Zaharia SE, Pietreanu CV (2018) Challenges in airport digital transformation. In: 7th International conference on air transport—INAIR (ScienceDirect, Transp Res Proc 35:90–99)
Zaharia SE, Pavel AP, Pietreanu CV (2020) Increasing employability in air transport trough new interdisciplinary qualifications. Proceedings of EduLearn20—12th annual international conference on education and new learning technologies. ISSN: 2340-1117

A Framework to Understand Current and Future Competences and Occupations in the Aviation Sector

Alessia Golfetti, Linda Napoletano, and Katarzyna Cichomska

Abstract Many of the major societal changes currently affecting the transportation system are expected to have a significant impact on jobs. Innovation and new technologies are redefining the human–machine partnership transforming the working environment. These macro changes are going to require new sets of skills, knowledge, and competences. Over the next years, it will be important to understand how these new technologies, innovations and procedures will change the job requirements. This article presents an analysis of occupations and competences required for current and emerging roles in the aviation sector. A mixed method approach was employed, which combined desk studies and the involvement of external aviation stakeholders.

Keywords Competences · Skills · Competence-based training · Future changes · Technological transformations · Changing occupations

1 Introduction

Many of the major societal changes currently affecting the transportation system are expected to have a significant impact on jobs. Over the next 10 years several global shifts, trends and innovations will shape the future of transportation, influencing working methods and practices. Together, technological, socio-economical, political, and demographic changes will generate new categories of jobs and occupations, while changing and displacing others (World Economic Forum 2016).

It is expected that by 2050, 68% of the global population will live in urban areas (Economic and social affairs of the United Nations 2018). Rising city populations and

A. Golfetti (✉) · L. Napoletano · K. Cichomska
Deep Blue s.r.l, Piazza Buenos Aires, 20, 00198 Rome, Italy
e-mail: alessia.golfetti@dblue.it

L. Napoletano
e-mail: linda.napoletano@dblue.it

K. Cichomska
e-mail: katarzyna.cichomska@dblue.it

M. Petrović and L. Novačko (eds.), *Transformation of Transportation*, EcoProduction,
https://doi.org/10.1007/978-3-030-66464-0_14

expanding metropolitan areas will require new transport solutions enabling smooth and well-connected journeys for people living both in the centre of the cities, as well as in their more remote areas (National Research Council 1996). Ageing and cultural diversity will further drive development of more inclusive transport means enabling elderly, disabled or otherwise vulnerable individuals to commute safely and comfortably, and offering those with language or cultural barriers a more accessible guidance and intuitive, international design (Britain 2008; OECD-Organisation for Economic Co-operation 2001; Hanson 2002; Design Council 2020; Bowering 2019; Shrestha et al. 2017).

Climate changes will also strengthen the need to develop more sustainable and environmentally friendly solutions facilitated by innovative ideas and technological advancements (National Research Council 1996; Arndt et al. 2013). These adjustments will drive not only changes to existing roles, but also create new jobs requiring entirely new sets of skills. With the globalisation on the rise, transversal, international skills, and more flexible labour conditions will also be needed (Glasbeek 2018). For the businesses to stay competitive, they must be aware of and understand the impact of these global trends on the working environment and resulting from them future labour requirements. Among other aspects, this will raise the importance of continuous, life-long education, dynamic on-the-job training, and ongoing re-qualification of the transportation workforce.

Digitalisation, automation, robotics, and decarbonisation represent the must-haves for the next 15 years and will impact the whole aeronautic sector. The amount of automation will increase for all roles transforming the roles of the operators and introducing new tasks that will require additional skills and capabilities. In the Air Traffic Management (ATM), the role of Air Traffic Controllers (ATCOs) will shift from an active and tactic-oriented role to one focused on monitoring and strategic decision-making (Doc 10056 2016). In Commercial Aviation, the concept of single pilot will replace the current concept of having two pilots in the cockpit. Augmented sensorial solutions and new coordination procedures and roles (e.g. the ground pilot) will support the pilot onboard in managing the flight (Bilimoria et al. 2014).

Digital analytics and AI will improve airline and airport operations. Robots and automation will help airports in handling a growing passenger number (International Air Transport Association 2018). Sustainable flying will be a requirement for both policy makers and end users, while Remotely Piloted Aircraft Systems (RPAS) will populate our urban areas due to a high number of applications (e.g. intelligent monitoring and surveillance, search and rescue, study and exploration, transport and delivery) (Undertaking 2017).

In this context, the fast-changing job market will entirely transform the nature of human-technology interaction in the workplace, requiring completely new sets of skills. To face these challenges, industries and educational bodies need to collaborate to make the school-to-work transition as smooth as possible, and to be able to effectively prepare future generations for the world of work (Lappas and Kourousis 2016). Lappas and Kourosis (2016) point out that this includes not only a consideration of the new type of skills that need to be taught, but also novel ways in which education is to be delivered in the first place to address the expectations and needs of the new generation of students.

As a result, the traditional educational system, characterised by generic training objectives and examination systems established according to the pre-defined level of knowledge expected from all participants, is now being transformed into a new system called Competency-Based Training (CBT) (Lappas and Kourousis 2016). CBT is based on specific needs and it is more learner-centred, allowing for training to be personalised and related more closely to the specific job in question and its tasks (Lappas and Kourousis 2016). More and more air transportation companies and transport service providers are adopting a competency-based such approach in their practices so that they can better align educational programmes and teaching methods to the needs of future operational scenarios and generations of workers. It has also been shown that applying CBT methodologies can lead to several benefits such as increased on-the-job performance, service proficiency and safety behaviours, all of which have a positive impact on customer satisfaction (Gibbs et al. 2017).

The study presented here was undertaken within the EU funded project entitled "Knowledge Alliance in the Air Transport" (KAAT)—KAAT project [2]. The KAAT project is an international, multi-partner project that aims to bridge the gap between the two pathways for education and training in the aviation sector: vocational and academic. It does that firstly by issuing a methodology for the Aviation Sectorial Qualification Framework, and secondly by modernising higher education through innovative approaches for teaching and learning. Through these activities, the project seeks to (1) ensure high quality future workforce, (2) foster innovation in teaching and facilitate the exchange and co-creation of knowledge, (3) ensure cohesive and transparent professional pathways, (4) enhance the international dimension of education and training in aviation, and (5) create a European collaboration network between the industry and education.

This article aims to present a competency framework and a sectorial breakdown of occupations developed to gain insights into the knowledge, skills and competences required by the current and emerging employers within the aviation sector.[1]

The article is structured as follows: Sect. 2 explains the competency framework, the sectorial breakdown of occupations and the methodological approach employed; Sect. 3 presents the main skills changes and emerging occupations expected in the sector; Sect. 4 addresses conclusions and future challenges for education and industries.

[1]This article is based on the KAAT Report on Occupational Analysis in Air Transport (2018) which can be accessed at https://www.kaat.upb.ro/wp-content/uploads/2018/12/KAAT_WP1_R1.1_Report-on-occupational-analysis_Final.pdf.

2 The Competency Framework and the Sectorial Breakdown of Occupations in the Aviation Sector

Within the KAAT project, an extensive analysis of current occupations and competences required by current jobs was carried out to better align educational programmes to the needs of future generations of workers. This included a two-stage approach comprising development of the sectorial breakdown of occupations (Sect. 2.1) and the competency framework (Sect. 2.2).

2.1 The Sectorial Breakdown of Occupations

The sectorial breakdown of occupations was developed to get a comprehensive view of the labour market in the aviation sector at EU level. It is a system of definitions describing the current occupations and related to them needs associated with the possession of specific knowledge, skills and competences by those holding the roles. The development of the sectorial breakdown has been conducted in compliance with the European classification of skills, competences and occupations (ESCO) with the aim for it to be used as a source for integrating and improving the labour market occupations in the aviation sector. The general structure of the sectorial breakdown is made up of two categories:

- Category 1 includes the eight high level areas of aviation activities identified according to the International Civil Aviation Organization (ICAO) classification of civil aviation activities (ICAO 2009) such as: Commercial aviation (Passenger and freight air transport operations with aircraft heavier than 5700 kg); General aviation (business aviation, instructional flying, aerial work, leisure flying); Airport services; Aerodrome services; Air navigation services; Regulatory functions; Other transportation support activities; Aviation training (not initial education but further training).
- Category 2 includes the 23 air professional areas describing different career paths related to the eight high level areas. For instance, for the commercial aviation category, the following professional areas have been identified: flight crew, cabin crew, other airline staff, commercial aircraft maintenance, and aircraft manufacturing.

These two levels constituted the basis for creating the current list of occupations in the aviation sector. Around 120 occupations were identified and fully described by providing definitions and descriptions of their key competences, tasks, responsibilities, skills, and knowledge. The European classification of Skills, Competences, Qualifications and Occupations (ESCO 2017) and relevant documentation at European level (EU) were used as initial sources for filling in the sectorial breakdown. The first outputs were then reviewed, validated, and expanded upon by a number of aviation stakeholders internal and external to the project (KAAT project 2018).

2.2 The Competency Framework

The competency framework was developed to describe the identified occupations included in the sectorial breakdown in a structured and consistent way. Competences, skills and tasks required to cover each role were identified, reviewed and grouped under meaningful categories that emerged from the analysis, providing a consistent approach to be used throughout the study. The competency framework is composed of eight categories representing "transversal" competences and "professional" competences (Fig. 1), including: (1) interpersonal skills and teamwork, (2) communication and reporting, (3) personal resilience and critical thinking, (4) training and development, (5) operational expertise, (6) customer focus, (7) leadership, management and planning, and (8) safety and responsibility. Each of these high-level categories of competence further includes a set of related to them key competences.

Each of the eight categories of competence contains: (1) an explanation of the key competences; (2) a list of keywords related to the supporting skills and personal qualities; and (3) a list of tasks and responsibilities that have been used for describing the identified occupations. Table 1 reports an example of the "interpersonal skills and teamwork" category of competence (KAAT project 2018). The complete list of competences can be accessed at https://www.kaat.upb.ro/wp-content/uploads/2018/12/KAAT_WP1_R1.1_Report-on-occupational-analysis_Final.pdf.

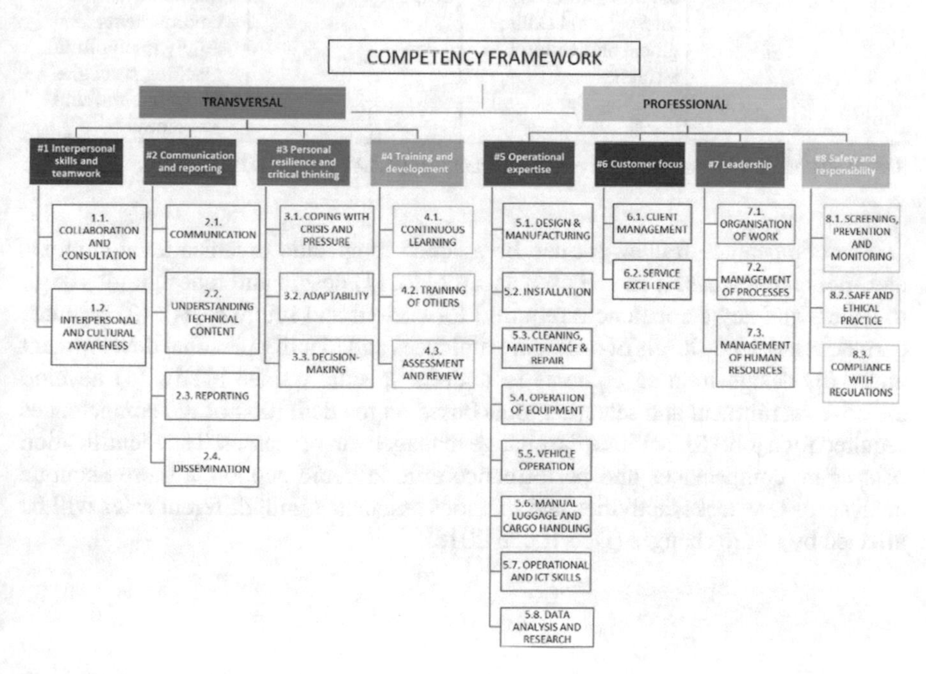

Fig. 1 Competency framework structure and classification. *Source* Report on occupational analysis in air transport [KAAT project 2018]

Table 1 Example of a description of competences: interpersonal skills and teamwork

Category of competence	Key competence	Supporting skills and personal qualities (Keywords)	Tasks and responsibilities
Transversal competence			
1. Interpersonal skills and teamwork	1.1 Collaboration and consultation Collaboration with other members of staff including joint completion of task; consultation with other professionals; smart distribution of tasks and consideration of individual strength and responsibilities; team leadership	Cooperation Consulting Facilitating teamwork Interacting Liaising Teamworking	• Complete work tasks as part of an aviation team • Consult with industry professionals • Cooperate with colleagues • Facilitate teamwork between students • Interact with airport stakeholder • Liaise with engineers • Work in an aviation team • Work in an logistic team
	1.2 Interpersonal and cultural awareness Interpersonal skills; global and cultural awareness	Having cultural awareness	• Ability to work in multicultural environments • Apply intercultural teaching strategies • Show intercultural awareness

Source Report on occupational analysis in air transport (KAAT project 2018)

The competency framework can be adapted to specific organisational contexts and application areas. It can be used as a tool to: (1) design and redesign jobs (e.g., to understand key competences required for current and future roles); (2) evaluate current competency levels of different employees and identify personal development areas; (3) design training contents to address specific trainee needs; (4) develop effective recruitment and selection tools based on the definition of the competences required for a job; (5) facilitate the change management processes. The identification of specific competences and performance criteria could support a more accurate analysis of how tasks, activities and methods associated with different roles will be affected by future changes (Doc 10056 2016).

2.3 Methodology

The sectorial breakdown and the competency framework for the aviation sector presented in this study were developed by using a combined methodological approach to identify and analyse: (1) current and emerging occupations; (2) key competences, responsibilities, skills and knowledge required for these occupations; and (3) current and future collaboration opportunities for industries and educational institutions.

The undertaken data collection comprised the following activities:

- a *top-down approach* focused on a structured review of current official documentation produced at a European (EU) level, and the analysis of past and ongoing EU funded projects.
  - The ESCO platform[2] and the review of EU sources was used to define the high-level categories and sub-categories of occupations developing the sectorial breakdown. Relevant documentation from EU reports, aviation documentation, and past and ongoing EU projects collected from project partners was used for identifying emerging and future occupations and enriching the descriptions of the current occupations.
  - The need to develop the competency framework emerged from the analysis of competences and tasks associated with the different roles identified, conducted based on ESCO classification. While undertaking the analysis, it was observed that the skills, competences, and tasks for each occupation were presented together under "essential/optional skills and competences" section of the website. It was felt necessary to make a distinction between these different elements to better align the descriptions of the occupations with the language of the labour market. Consequently, several relevant competency frameworks were identified through an Internet search and reviewed. The European Union Safety Agency (EASA) framework (EASA report 2015) was selected as most applicable and modified to better align the key categories of competences with the tasks and responsibilities identified through the ESCO website. The framework was then enlarged by adding missing competences; thus, creating a new competency framework for occupations in the aviation sector.
  - As a final step, tasks, responsibilities, and skills/personal qualities associated with the identified occupations were mapped onto the categories of competences forming the framework. This step was done with the help of project partners and external experts.
- A *bottom-up approach* where the involvement of external aviation stakeholders was a key element in evaluating and validating the sectorial breakdown and the competency framework. Two different activities were organised:
  - The administration of an online survey for collecting feedback regarding current and future occupations and competences in the aviation sector. The purpose of the online survey was to gather input from employees working in

[2]The ESCO platform can be accessed at https://ec.europa.eu/esco/portal/home.

all areas of the aviation sector value chain including: Commercial Aviation, General Aviation, Airport Operations, Air Navigation Services, Regulatory Functions, Other Transportation Support Activities, and Aviation Training. The survey was open for a period of 3 months, from 15.05.18 to 31.07.18. During this time, a total of 132 responses were received.

– A workshop on "Smart qualifications for smart air transport occupations" organised in Lisbon in July 2018. Experts representing different areas within the aviation sector participated in the workshop to validate the classification of competences and the structure of the sectorial breakdown of occupations.

3 How Will the Importance of Different Roles and Competences Change in the Next Years? Main Findings from the Survey

The impact of new technologies such as robotics, autonomous vehicles and big data is transforming the working environment, changing the skills that employees need for doing their job. This trend was also highlighted in the results gathered through the survey forming part of this study. Among other aspects, the survey aimed to identify the occupational needs in the aviation sector in the next 10 years. Most of the respondents reported that information technologies, cooperative systems, big data and augmented reality interfaces will heavily affect the current occupations and the way of working, requiring new knowledge, skills and abilities (Fig. 2).

As depicted by the World Economic Forum (2016), the twenty-first century skills are mostly related to problem solving, critical thinking and creativity. The Future of Jobs Report 2018 (World Economic Forum 2018) reported that future workers generations will need to be equipped with a plethora of skills, like the ability to respond to complex problems, effective communication and team working. These findings are also in line with the results gathered from the online survey (Fig. 3).

Most respondents indicated dealing with complexity, critical thinking, communication and reporting, and teamwork and collaboration as not only the essential competences required even more in the near future, but also critical success factors for the future careers. At the same time, the importance of competences such as teaching, advising and coaching and technical expertise will be less required in the coming years compared to the other investigated competences. One of the reasons why this may be the case is due to the use of new intelligent systems that will be able to provide ongoing guidance on operations and ad hoc technical support facilitated by advanced technologies such as augmented and virtual reality, as a result requiring less technical and operational knowledge from the humans themselves.

As reported in a recent Airbus publication (Airbus Employment Marketing 2018), and in line with the growing importance of the cross functional skills, the aviation engineer of the future will require a mixture of technical and soft skills that are related to the current context of digitalisation and increasingly rapid technological change, including:

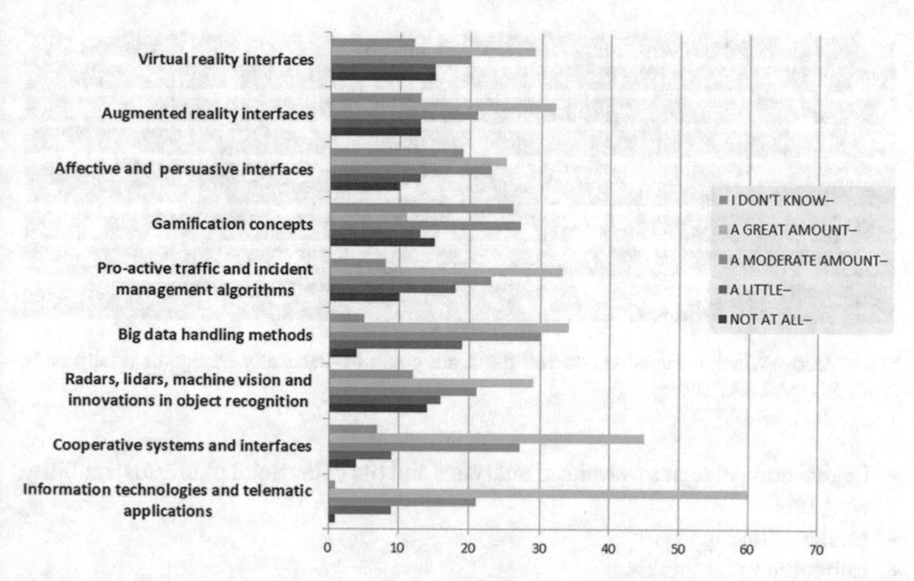

Fig. 2 Q29—In your opinion, to what degree will the following changes and key technologies affect your current occupation in the aviation sector? *Source* KAAT project survey (KAAT project 2018)

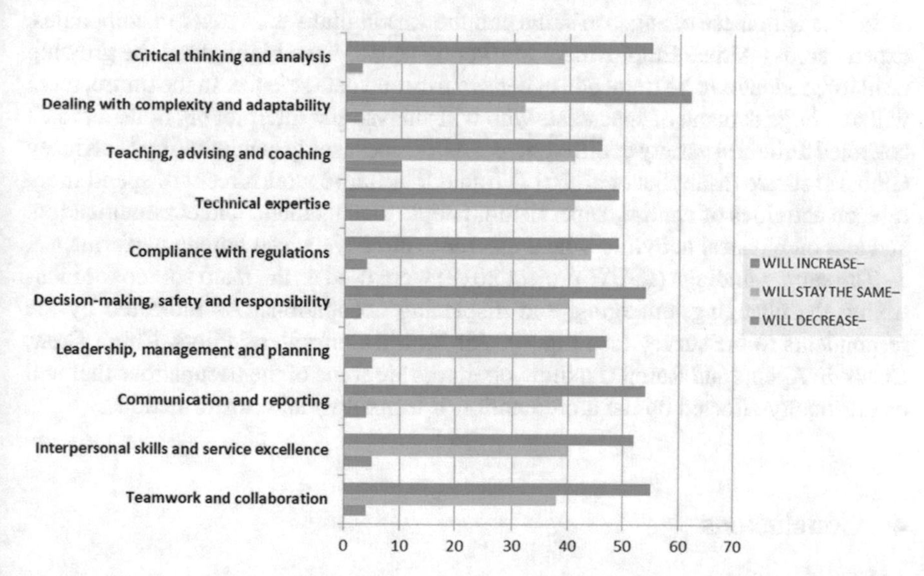

Fig. 3 Q26—In the next 10 years, do you think the importance of these competences will stay the same, increase or decrease? *Source* KAAT survey (KAAT project 2018)

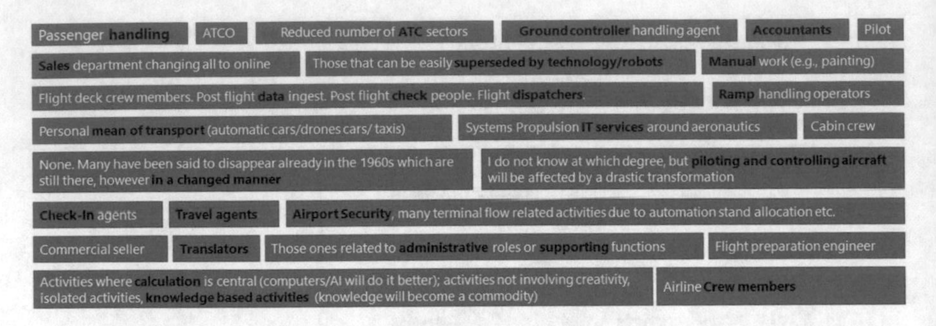

Fig. 4 Q30—Which occupations do you think are going to drastically change or disappear by 2030? *Source* KAAT survey

- Digital competencies (advanced analytics and big data, cloud platforms, mobility, etc.)
- Design thinking
- Entrepreneurial thinking
- Cyber security skills
- Skills related to virtual/augmented reality.

In line with these changes in skills and the vast digitalisation trend in companies, experts across Airbus Employment Marketing (2018) have highlighted the growing need for graduates to be trained in cyber security and data science. In the future, there will be a huge demand of specialists who will analyse and interpret big transport data collected through a variety of different means. A recent study conducted by McKinsey Global Institute (Manyika et al. 2017) found that future workforce will spend more time on activities of control, supervision, people management, and communication, and less on physical activities where machines already exceed human performance.

The survey findings (KAAT project 2018) were used as the main source for identifying the changing, emerging, and displacing occupations. As indicated by the respondents to the survey (see Fig. 4), Air Traffic Controllers, Pilots, Cabin Crew, Check-in Agents and Ramp Handling operators are some of the occupations that will be drastically affected by the aforementioned technological transformations.

4 Conclusions

The aviation sector depends on highly skilled employees whether they are pilots, engineers, air traffic controllers, safety inspectors, or representatives of other roles. The digital revolution is transforming the aviation industry demanding skills that previously were not necessary.

To meet the new learning needs, teachers need to move away from traditional teaching methods and adopt educational approaches that are more in line with the way students learn today and the new demands of the workforce of tomorrow (Lappas and Kourousis 2016). This result highlights the need to foster collaboration between organisations and educational institutions and to ensure an involvement of key stakeholders from industries and universities in order to allow educational programs to be effective and up-to-date with the real issues required by the world of work. Furthermore, a major emphasis on professional and competency-based training will be needed.

On the other hand, in order to make current and future workforce ready to meet the labour market requirements, industries need to (1) identify productive ways of planning job transitions pathways; (2) invest in continual upskilling and upgrading of competencies of the workforce; (3) recognise and understand future skills demand; (4) re-design training courses to foster continued learning, and (5) promote on-the-job training opportunities to facilitate chances to acquire new skills in the workplace (World Economic Forum 2017).

The recent months have furthermore shown how an unexpected worldwide crisis, caused in this case by COVID-19, can very suddenly cause major disruption not only to the working lives of billions of people, but also to their everyday private lives, entirely destabilising otherwise fairly solid market. The state of the pandemic has over a matter of days revolutionised entirely the way people work, moving regular offices to people's private homes, closing many services that were relied upon in everyday life like shops and restaurants, and suspending travel both locally and across the countries. The aviation sector itself experienced a huge drop of income caused by the closures of the borders and fear of the spread of the disease resulting in cancellations of millions of flights and connections across the globe. In Italy alone, during the first weeks of country-wide lockdown in March 2020, the number of flights dropped by up to 98% comparing to the same time in the previous year and has only partially recovered since (Statista Research Department 2020). The passenger revenue loss is currently estimated at 314 billion U.S. dollars globally just for the airlines themselves (Mazareanu 2020). Consequently, this has led to significant number of pay cuts and layoffs.

These sudden, drastic steps further emphasise the need for organisations and workers to be resilient and adaptable, both in terms of the services they offer, the skills they possess, and the way and pace in which they work (De Smet et al. 2020). This has been visible by the quick adaptations seen, for example, within the manufacturing sector where almost overnight companies have changed their production lines to help the widespread fight against the disease, and to help replenish the very quickly stressed and much needed resources such as hand sanitisers (e.g., Gin58, LVMH), surgical masks and visors (e.g., Sharp, Royal Mint, Fred), protective uniforms (e.g., Louis Vuitton, Christian Dior), diagnostic equipment (e.g., Bosch), and breathing apparatuses (e.g., Dyson).

People and organisations had to adapt to those changing demands almost instantly, and those workplaces with more flexible working practices and wider array of skills and knowledge were not only affected less dramatically and able to keep their businesses running, but some of them were also capable of designing, developing or producing necessary products, providing free guidance and expertise, or ensuring the wellbeing of staff and the wider community by offering free psychological support.

Many workplaces, factories, and large spaces such as sport stadiums, exposition halls or car parks have been repurposed as people mobilised to address shortages in critical supplies and patient facilities. Although not without its challenges and stumbling blocks, this fight continues, and further resources and guidance materials have been produced to help organisations become more adaptable, resilient, and better prepared for facing such critical situations (López-Gómez et al. 2020).

This study has presented an analysis of the occupations in the aviation sector conducted within the KAAT project. The competency framework and the sectorial breakdown of occupations were presented as tools for supporting training providers, educational institutions and industries in understanding the key competences required for current and future roles, and designing new training pathways for skilling, upskilling and reskilling of the aviation workforce.

The competency framework has the potential value to be adapted to specific operational contexts and organisational requirements. It could also be used to identify those skills and knowledge areas which have been in strong demand during the pandemic, and to review the condition of the aviation sector following the crisis. It is without a doubt that COVID-19 has introduced some irreversible changes to the working lives of many people and, like with any situation requiring fast and widespread adaptation, it is important to review and learn from these experiences to build a stronger and more resilient working environment of the future.

As the next step, the framework can be further improved and extended by taking into consideration other relevant methodologies adopted at the EU level, through further validation of the existing content, as well as application and customisation of the framework to other industry sectors.

Acknowledgements This work is co-funded by the Erasmus + programme of the European Commission through the projects: "Knowledge Alliance in Air Transport (KAAT)", project N° 588060-EPP-1-2017-1-RO-EPPKA2-KA and "skilling, upskilling and reskilling in the future Air Transport (skill-UP)", project N° 408540-EPP-1-2019-1-IT-EPPKA2-SSA. The authors wish to express their gratitude to project partners and to the experts involved in the study.

References

Airbus Employment Marketing (2018) The engineer of the future. White paper, 2018. https://company.airbus.com/careers/Partnerships-and-Competitions/The-Engineer-of-the-Future-White-Paper.html

Arndt WH, Schäfer T, Emberger G, Tomaschek J (2013) Transport in megacities—development of sustainable transportation systems. https://www.researchgate.net/publication/273996052_Transport_in_Megacities_-development_of_sustainable_transportation_systems

Bilimoria KD, Johnson WW, Schutte (2014) Conceptual framework for single pilot operations. In: Proceedings of the international conference on human-computer interaction in aerospace, pp 1–8

Bowering T (2019) Ageing, mobility and the city: objects, infrastructures and practices in everyday assemblages of civic spaces in East London. J Popul Ageing 12(2):151–177 https://link.springer.com/article/10.1007/s12062-019-9240-3

Britain G (2008) Inclusion by design: equality, diversity and the built environment. Commission for Architecture and the Built Environment. https://www.designcouncil.org.uk/sites/default/files/asset/document/inclusion-by-design.pdf. Accessed 5 Aug 2020

Design Council (2020) https://www.designcouncil.org.uk/news-opinion/designing-diversity

De Smet A, Pacthod D, Relyea C and Sternfels B (2020) Ready set, go: Reinventing the organization for speed in the post-COVID-19 era. McKinsey & Company, 26 June 2020. https://www.mckinsey.com/business-functions/organization/our-insights/ready-set-go-reinventing-the-organization-for-speed-in-the-post-covid-19-era. Accessed 17 July 2020

Doc ID (2016) 10056 manual on air traffic controller competency-based training and assessment first edition—2016

EASA Report (2015) Proposal for a competency framework for the competent authorities' inspectors. https://www.easa.europa.eu/sites/default/files/dfu/EASA%20Aviation%20Inspector%20Competencies%20Report.pdf

Economic and social affairs United Nations (2018) World urbanisation prospects: The 2018 Revision. https://population.un.org/wup/Publications/Files/WUP2018-KeyFacts.pdf. Accessed 5 Aug 2020

Erasmus+ Knowledge Alliance in Air Transport (KAAT) Project (2017) (588060-EPP-1-2017-1-RO-EPPKA2-KA). https://www.kaat.upb.ro/

Erasmus+ Knowledge Alliance in Air Transport (KAAT) Project, (588060-EPP-1-2017-1-RO-EPPKA2-KA), R1.1. Report on occupational analysis in air transport (2018). https://www.kaat.upb.ro/wp-content/uploads/2018/12/KAAT_WP1_R1.1_Report-on-occupational-analysis_Final.pdf

Gibbs L, Slevitch L, Washburn I (2017) Competency-based training in aviation: The impact on flight attendant performance and passenger satisfaction. J Aviat/Aerosp Educ Res 26(2):55–80. https://commons.erau.edu/cgi/viewcontent.cgi?article=1716&context=jaaer. Accessed 5 Aug 2020

Glasbeek S (2018) The importance of transversal skills and competences for young people in a modern Europe. Policy Paper by The Youth Development Working Group by AEGEE Europe https://www.aegee.org/policy-paper-the-importance-of-transversal-skills-and-competences-for-young-people-in-a-modern-europe/. Accessed 5 Aug 2020

Handbook ES (2017) European skills, competences, qualifications and occupations (2017) EC Directorate E. https://ec.europa.eu/esco/portal/home

Hanson J (2002) The inclusive city, what active ageing might mean for urban design. https://discovery.ucl.ac.uk/id/eprint/3319/1/3319.pdf. Accessed 5 Aug 2020

ICAO working paper (2009) Review of the classification and definitions used for civil aviation activities. https://www.icao.int/Meetings/STA10/Documents/Sta10_Wp007_en.pdf

International Air Transport Association (2018) Future of the airline industry 2035. https://www.iata.org/policy/Documents/iata-future-airline-industry.pdf

Lappas I, Kourousis KI (2016) Anticipating the need for new skills for the future aerospace and aviation professionals. J Aerosp Technol Manage 8(2):232–241

López-Gómez C, Corsini L, Leal-Ayala D, Fokeer S (2020) COVID-19 critical supplies: the manufacturing repurposing challenge. United Nations Industrial Development Organization (UNIDO), April 2020. https://www.unido.org/news/covid-19-critical-supplies-manufacturing-repurposing-challenge. Accessed 17 July 2020

Manyika J, Lund S, Chui M, Bughin J, Woetzel J, Batra P, Ko R, Sanghvi S (2017) Jobs lost, jobs gained: workforce transitions in a time of automation. McKinsey Global Institute, 150. https://www.mckinsey.com/~/media/mckinsey/featured%20insights/future%20of%20organizations/what%20the%20future%20of%20work%20will%20mean%20for%20jobs%20skills%20and%20wages/mgi-jobs-lost-jobs-gained-report-december-6-2017.ashx

Mazareanu E (2020) Coronavirus: impact on the aviation industry worldwide—statistics and facts. Statista. Published on 26 June 2020. https://www.statista.com/topics/6178/coronavirus-impact-on-the-aviation-industry-worldwide/. Accessed 17 July 2020

National Research Council (1996) Transportation options for megacities in the developing world. In: Meeting the challenges of megacities in the developing world: a collection of working papers (pp 1–4). https://www.nap.edu/read/5267/chapter/4. Accessed 5 Aug 2020

OECD-Organisation for Economic Co-operation (2001) Ageing and transport: mobility needs and safety issues. Organization for Economic. https://www.oecd-ilibrary.org/docserver/9789264195851-sum-en.pdf?expires=1596626665&id=id&accname=guest&checksum=AE7A2A06AD3002EA859B08CDA473FDB4. Accessed 5 Aug 2020

Shrestha BP, Millonig A, Hounsell NB, Mcdonald M (2017) Review of public transport needs of older people in European context. J Popul Ageing 10(4):343–361. https://link.springer.com/article/10.1007/s12062-016-9168-9

Statista Research Department (2020) Year-on-year change in daily airport passenger traffic due to the coronavirus outbreak in Italy from February 24 to March 22, 2020. Published on 18 June 2020. https://www.statista.com/statistics/1107036/coronavirus-impact-airport-passenger-traffic-italy/. Accessed 17 July 2020

Undertaking SJ (2017) U-space blueprint. SESAR Joint Undertaking. Accessed 18 Sept. https://www.sesarju.eu/sites/default/files/documents/reports/U-space%20Blueprint%20brochure%20final.PDF. Accessed 17 July 2020

World Economic Forum (2016) The future of jobs: employment, skills and workforce strategy for the fourth industrial revolution. In: Global challenge insight report. World Economic Forum, Geneva. https://www.weforum.org/docs/WEF_Future_of_Jobs.pdf. Accessed 17 July 2020

World Economic Forum (2017) Accelerating workforce reskilling for the fourth industrial revolution: an agenda for leaders to shape the future of education, Gender and Work. World Economic Forum, Geneva, Switzerland. https://www.weforum.org/docs/WEF_EGW_White_Paper_Reskilling.pdf. Accessed 17 July 2020

World Economic Forum (2018) The future of jobs report 2018. World Economic Forum, Geneva. https://www.weforum.org/docs/WEF_Future_of_Jobs_2018.pdf. Accessed 17 July 2020

Printed in the United States
by Baker & Taylor Publisher Services